高等院校观赏园艺方向"十二五"规划教材

插花艺术双语教程
（中英）

郑　丽　陈利平　主编

中国林业出版社

图书在版编目(CIP)数据

插花艺术双语教程：汉英对照 / 郑丽，陈利平主编. —北京：中国林业出版社，2015.12（2025.1重印）
高等院校观赏园艺方向"十二五"规划教材
ISBN 978-7-5038-8307-1

Ⅰ. ①插… Ⅱ. ①郑… ②陈… Ⅲ. ①插花－装饰美术－高等学校－教材－汉、英 Ⅳ. ①J525.1
中国版本图书馆CIP数据核字(2015)第300135号

中国林业出版社·教育出版分社
策划编辑：康红梅　　　　　责任编辑：田　苗
电话：83143557　　　　　　传真：83143516

出版发行	中国林业出版社(100009　北京市西城区德内大街刘海胡同7号) E-mail: jiaocaipublic@163.com　电话：(010) 83143500 http://lycb.forestry.gov.cn/
经　销	新华书店
印　刷	北京中科印刷有限公司
版　次	2015年12月第1版
印　次	2025年1月第2次印刷
开　本	850mm×1168mm　1/16
印　张	9
彩　插	1印张
字　数	232千字
定　价	58.00元

未经许可，不得以任何方式复制或抄袭本书之部分或全部内容。
版权所有　侵权必究

高等院校观赏园艺方向规划教材
编写指导委员会

顾　问　陈俊愉（北京林业大学）
主　任　张启翔（北京林业大学）
副主任　李　雄（北京林业大学）
　　　　　　包满珠（华中农业大学）
　　　　　　李树华（清华大学）
委　员（按姓氏拼音排序）
　　　　　　包志毅（浙江农林大学）
　　　　　　车代弟（东北农业大学）
　　　　　　陈发棣（南京农业大学）
　　　　　　高俊平（中国农业大学）
　　　　　　高亦珂（北京林业大学）
　　　　　　何少云（华南农业大学）
　　　　　　何松林（河南农业大学）
　　　　　　蒋细旺（江汉大学）
　　　　　　金研铭（吉林农业大学）
　　　　　　亢秀萍（山西农业大学）
　　　　　　吴少华（福建农林大学）
　　　　　　姚允聪（北京农学院）
　　　　　　于晓英（湖南农业大学）
　　　　　　岳　桦（东北林业大学）
　　　　　　曾　明（西南大学）
　　　　　　张　钢（河北农业大学）
　　　　　　郑成淑（山东农业大学）
秘书长　高亦珂（北京林业大学）
　　　　　　康红梅（中国林业出版社）

编 写 人 员

主　　编　郑　丽　陈利平

副 主 编　彭春秀　盛爱武　吴玉美

编译人员　（按姓氏拼音排序）

包晓鹏（云南省楚雄州农业科学研究推广所）

陈鸿霖（云南农业大学）

陈利平（云南农业大学）

陈友华（云南农业大学）

段国珍（昆明花语花艺插花培训学校）

高笑梅（云南农业大学）

林　燕（昆明学院）

毛　瑾（云南农业大学）

彭春秀（云南农业大学）

盛爱武（仲恺农业工程学院）

王海燕（昆明学院）

吴广平（云南农业大学）

吴玉美（昆明学院）

郑　丽（云南农业大学）

周　雯（云南农业大学）

主　　审　资谷生　Marcia Eames-Sheavly

前言

我国享有"世界园林之母"的美誉,花卉与我国悠久而灿烂的历史文化有着密切的关系。古人云:养花,修身养性也。自古以来插花与闻香、品茗、挂画并称为"四般闲事",这些都是中国人怡情养性的优雅生活;美国出版的花卉园艺畅销书把中国的一句谚语"种植鲜花的人也种植了快乐"置于扉页;英国人提及花卉时说到:"没有中国花卉的园子不能称之为花园"……足见,中国花卉及其文化在世界范围内的地位是相当高的。

随着人类文明的不断进步,插花、闻香、品茗、挂画这些源于中国的生活艺术被越来越广泛地传播到世界各地,并被传承与弘扬。特别是针对目前越来越快的生活节奏,人类已经很难"闲"下来,这"四般闲事"已然成了现代人渴望健康、向往修养的生活方式。如在与我们隔海相望的日本,花道、茶道、香道不仅成为了他们的国粹,更成为从上流社会到市民阶层都乐于接受的修身养心之举。而这些中华民族绚丽多彩的典雅生活艺术,仅就"插花"一项,对于今天的国人而言,都陌生到成为束之高阁的专业领域,以至于在国际层面上,日本花道的影响力已远胜于我国的插花艺术。

所幸,近年来云南省已建设成为我国"花卉大省"和面向东南亚的"桥头堡",在这里各类国内外的"花事"不断,"插花"这门看似专业的艺术再次回归到百姓的生活中。越来越多的人,特别是东南亚的国际友人希望到花之都——云南来学习插花;不少专业学子基于职业的规划也希望通过学习获得"插花员"资格。我们望通过《插花艺术双语教程(中英)》的推出,把我国优秀灿烂的插花艺术文化介绍给更多的国际友人,为进一步巩固我国"世界园林之母"的地位有所贡献。

考虑到教材的针对性和实用性,本教材以人力资源和社会保障部教材办公室组织编撰出版的《插花员教材》为蓝本,力求精简浓缩,以插花艺术基础知识为纲要,旨在系统且深入浅出地介绍插花艺术,以适合更多专业或非专业人士阅读。全书共分为6章,第1章插花艺术概论,主要介绍插花艺术的定义、历史、分类和作用;第2章简要介绍艺术插花的花材和花器;第3章介绍常见的礼仪插花;第4章通过作品分析重点阐述花文化内涵与插花的关系,这是较以往插花教材的创新部分;第5章主要介绍插花制作基础及常用手法;第6章主要介绍花材保鲜知识和花卉包装材料应用知识。另外,为方便读者查询和学习,将重点的专业词汇归纳附于书后。

本教材由中国花卉协会花文化专业委员会常务理事、云南农业大学园林园艺学院郑丽以及外语学院陈利平担任主编。具体编写分工为：第 1 章由郑丽、吴玉美编写，陈鸿霖、毛瑾、高笑梅、吴广平翻译；第 2 章由盛爱武、吴玉美、包晓鹏编写，陈鸿霖、毛瑾、吴广平翻译；第 3 章由林燕、周雯、段国珍、王海燕编写，陈鸿霖、毛瑾、吴广平翻译；第 4 章由郑丽、盛爱武编写，高笑梅翻译；第 5 章由林燕、周雯、段国珍编写，吴广平、陈鸿霖、毛瑾翻译；第 6 章由彭春秀编写，陈友华翻译。尤其要感谢云南农业大学盛军校长对本教材的关心与支持。特别感谢田松青、王梓然、洪明伟、杨发顺、艾万峰为本教材提供图片。本教材由云南农业大学外语学院院长资谷生教授、美国康奈尔大学资深花艺学者 Marcia Eames-Sheavly 审定。

《插花艺术双语教程（中英）》适用于园艺专业、英语翻译专业及国际学生的教学。同时也可供插花和英语翻译爱好者阅读参考。由于编者水平有限，书中错误难免，望广大读者批评指正！

郑 丽
2015 年 4 月于美国田纳西大学

Preface

China enjoys the reputation of "the Mother of All Gardens in the World", and flowers have a close relationship with China's long history and splendid civilization. As the ancients said, "Flower growing is an activity to cultivate one's moral character and mould one's temperament." Since ancient times, flower arrangement, incense smelling, tea tasting and painting hanging have been known as the "four leisurely activities". Actually they represent Chinese people's elegant life in pursuit of the inner tranquility. A horticultural bestseller published in America carries a Chinese proverb on its cover — "One who grows flowers also grows happiness". Speaking of flowers, British people always say, "A garden without Chinese flowers is not a garden". It shows that Chinese flowers and their culture are of a remarkably high status in the world.

With the constant development of human civilization, the arts of life originated in China, such as flower arrangement, incense smelling, tea tasting and painting hanging, have been becoming more and more popular around the world, and inherited and carried forward as well. Facing the stressful and busy life, people at present feel it hard to enjoy a moment of leisure. The "four leisurely activities" have become their philosophy of life to show their yearning for a healthy and relaxing life. In Japan, our neighboring country, ikebana (flower arrangement), tea ceremony and incense ceremony have become not only the quintessence of Japanese culture, but also the activities embraced by people of all social strata. However, among these wonderful and splendid Chinese arts, flower arrangement is so strange to most Chinese people at present that it has become an area of expertise. The influence of Japanese ikebana has surpassed that of Chinese flower arrangement at the international level.

Fortunately, in recent years, as Yunnan province has being built into a major province of flower industry and the "bridgehead" of Southeast Asia, various national and international "flower events" were constantly held, and flower arrangement, a seemingly professional art, returned to the life of ordinary people. More and more people, especially those from Southeast Asia, hope to learn flower arrangement in Yunnan — the homeland of flowers. Many students majoring in horticulture also want to gain the qualification for flower arrangers. By publishing *Floral Art Bilingual Course (English-Chinese Edition)*, we hope to introduce the wonderful Chinese flower arrangement art to foreign friends, and help to further consolidate China's position as the "Mother of All Gardens in the World".

To make it less specialized and more practical, this book is written with *A Textbook for Flower Arrangers* edited and published by the Textbook Office of the Ministry of Human Services and Social Security as the blueprint. Focusing on the basic knowledge of flower arrangement, the book is aimed at introducing flower arrangement art in simple and concise language to make it readable for more professionals or nonprofessionals. The book is composed of six chapters. Chapter 1, a general introduction to flower arrangement art, gives a brief introduction to the definition, history, classification and function. Chapter 2 briefly introduces flower materials and conventional equipment for artistic flower arrangement. Chapter 3 introduces common flower arrangement for etiquette. Chapter 4 puts emphasis on the relationship between the connotation of flower culture and artistic flower arrangement through the analysis of specific works. It is also an innovative part of this book compared with the previous flower arrangement textbooks. Chapter 5 mainly introduces the basis of production of flower arrangement and the commonly used techniques. Chapter 6 introduce the fundamentals for preserving cut flowers and applying flower packaging material respectively. In addition, the key terms are attached both in Chinese and English for readers to study and consult.

The chief editor of the Chinese version of the book is Zheng Li, Professor of the college of Landscape of Yunnan Agricultural University, and executive member of Chinese Association of Flower Civilization, while Chen Liping, professor of the college of Foreign Languages, is the chief editor of the English version. We are particularly grateful to Mr. Sheng Jun, president of Yunnan Agricultural University, for his concern and support for this book. The following are the specific writers and translators of each chapter: Chapter One is written by Zheng Li, Wu Yumei, and translated by Chen Honglin, Mao Jin, Gao Xiaomei and Wu Guangping. Chapter Two is written by Sheng Aiwu, Wu Yumei, Xiaopeng, and translated by Chen Honglin, Mao Jin and Wu Guangping. Chapter Three is written by Lin Yan, Zhou Wen, Duan Guozhen,

Wang Haiyan and translated by Chen Honglin, Mao Jin and Wu Guangping. Chapter Four is written by Zheng Li, Sheng Aiwu, and translated by Gao Xiaomei. Chapter Five is written by Lin Yan, Zhou Wen, Duan Guozhen and translated by Wu Guangping, Chen Honglin, and Mao Jin. Chapter Six is written by Peng Chunxiu, and translated by Chen Youhua. Special thanks also goes to Tian Songqing, Wang Ziran, Hong Mingwei, Yang Fashun and Ai Wanfeng for providing photos of the flower arrangement works used in this book. The whole book is reviewed and improved by Professor Zi Gusheng, Dean of the College of Foreign Languages of Yunnan Agricultural University, and Marcia Eames-Sheavly, senior floriculture scholar of Cornell University.

Flower Arrangement Art (*English-Chinese Edition*) can be used to teach students majoring in horticulture and English translation, as well as international students. It can also serve as a reading and referential material for those who are interested in flower arrangement art and English translation. Due to the limited competence of the editors, there may be some mistakes in the book inevitably, and we do hope our readers kindly point out the mistakes for us to correct in the next edition.

Zheng Li
April, 2015 at the University of Tennessee, U.S.

前言

第1章 插花艺术概述

1.1 插花艺术的定义 ………………………………………… 2
1.2 插花艺术的历史 ………………………………………… 3
1.2.1 中国插花历史简介 ………………………………… 3
1.2.2 日本插花历史简介 ………………………………… 12
1.2.3 西方插花历史简介 ………………………………… 14
1.3 插花艺术的分类 ………………………………………… 15
1.3.1 按民族风格分类 …………………………………… 15
1.3.2 按时代分类 ………………………………………… 15
1.3.3 按器具分类 ………………………………………… 19
1.3.4 按表现手法分类 …………………………………… 20
1.3.5 按使用场合分类 …………………………………… 21
1.4 插花艺术的作用 ………………………………………… 22
1.4.1 环境作用 …………………………………………… 22
1.4.2 文化作用 …………………………………………… 22
1.4.3 感情作用 …………………………………………… 23
1.4.4 经济作用 …………………………………………… 23
1.5 插花艺术评判标准 ……………………………………… 24
1.5.1 构图造型 …………………………………………… 24
1.5.2 色彩配置 …………………………………………… 24
1.5.3 主题意境 …………………………………………… 24
1.5.4 制作技巧 …………………………………………… 25

第2章 花材花器

2.1 花卉的色彩和形状 ……………………………………… 28
2.1.1 花卉的色彩 ………………………………………… 28
2.1.2 花卉的形状 ………………………………………… 30

2.2 常用草本花材及花语 …………………………………………… 31

2.3 常用木本花材 …………………………………………………… 31

2.4 常用插花器具 …………………………………………………… 39

    2.4.1 常用插花器皿 …………………………………………… 39

    2.4.2 常用工具和辅件 ………………………………………… 40

第3章 礼仪用花

3.1 婚庆用花 ………………………………………………………… 46

    3.1.1 新娘花 …………………………………………………… 46

    3.1.2 花车 ……………………………………………………… 48

    3.1.3 胸花 ……………………………………………………… 51

    3.1.4 撒花 ……………………………………………………… 52

3.2 庆典用花 ………………………………………………………… 52

    3.2.1 制作特点 ………………………………………………… 52

    3.2.2 庆典花篮 ………………………………………………… 54

3.3 丧事用花 ………………………………………………………… 55

    3.3.1 花圈 ……………………………………………………… 56

    3.3.2 花篮花束 ………………………………………………… 56

    3.3.3 摆花 ……………………………………………………… 57

3.4 会议和宴会用花 ………………………………………………… 57

    3.4.1 制作特点 ………………………………………………… 57

    3.4.2 常用形式 ………………………………………………… 58

3.5 生日用花 ………………………………………………………… 59

    3.5.1 制作特点 ………………………………………………… 59

    3.5.2 常用形式 ………………………………………………… 61

3.6 探望病人用花 …………………………………………………… 62

    3.6.1 制作特点 ………………………………………………… 62

    3.6.2 常用形式 ………………………………………………… 62

3.7 其他场合用花 …………………………………………………… 62

第4章　插花命题及花文化内涵

4.1 插花的命题立意 ·· 66
　　4.1.1　命题立意的特点 ·· 66
　　4.1.2　命题立意的要求 ·· 66
　　4.1.3　命题立意的基本方法 ·· 67
　　4.1.4　命题立意的具体方法 ·· 67
　　4.1.5　提高命题立意素养的方法 ·· 68

4.2 插花的欣赏与花文化 ·· 69
　　4.2.1　花卉欣赏与花文化 ··· 70
　　4.2.2　花卉欣赏要点 ··· 71
　　4.2.3　花卉文化内涵在作品中的表现 ······································ 72

第5章　插花制作基础及常用手法

5.1 花艺制作技法概述 ·· 82
　　5.1.1　花艺材料加工 ··· 82
　　5.1.2　花艺结构构成 ··· 83

5.2 常用表现手法 ··· 85
　　5.2.1　立体构成法 ·· 85
　　5.2.2　一枝突出法 ·· 87
　　5.2.3　斜角呼应法 ·· 88
　　5.2.4　分组组合法 ·· 89

5.3 插花基本技巧 ··· 91
　　5.3.1　准备工作 ·· 91
　　5.3.2　修理花材 ·· 91
　　5.3.3　花材固定 ·· 94
　　5.3.4　作品整理 ·· 95

第6章　花材保鲜及包装

6.1 花材保鲜 ·· 98
　　6.1.1　花材萎蔫的原因 ·· 98
　　6.1.2　花材的保鲜原理 ·· 101
　　6.1.3　花材的保鲜方法 ·· 103

6.1.4 花卉保鲜剂 …………………………………………………… 106

6.2 花卉包装 ………………………………………………………… 112
    6.2.1 花卉包装与顾客心理 …………………………………… 112
    6.2.2 现代包装材料 …………………………………………… 116

参考文献 ……………………………………………………………… 123

附录 插花艺术术语英汉对照表 …………………………………… 124

彩图 …………………………………………………………………… 127

Contents

Preface

Chapter 1 A General Introduction to Flower Arrangement Art

- *1.1* Definition of Flower Arrangement Art 2
- *1.2* History of Art of Flower Arrangement Art 3
  - 1.2.1 A Brief History of Chinese Flower Arrangement 3
  - 1.2.2 A Brief History of Japanese Flower Arrangement 12
  - 1.2.3 A Brief history of Occidental Flower Arrangement 14
- *1.3* Classification of Flower Arrangement Art 15
  - 1.3.1 Classification According to National Styles 15
  - 1.3.2 Classification According to Ages 15
  - 1.3.3 Classification According to Equipment 19
  - 1.3.4 Classification according to Expression Technique 20
  - 1.3.5 Classification According to Application 21
- *1.4* Functions of Art of Flower Arrangement 22
  - 1.4.1 Environmental Function 22
  - 1.4.2 Cultural Function 22
  - 1.4.3 Emotional Function 23
  - 1.4.4 Economic Function 23
- *1.5* Criteria for Appraising Flower Arrangement Art 24
  - 1.5.1 Composition and Plastic Art 24
  - 1.5.2 Color Scheme 24
  - 1.5.3 Theme of Artistic Conception 24
  - 1.5.4 Producing Skills 25

Chapter 2 Flower Materials and Conventional Equipment for Flower Arrangement

- *2.1* Color and Shape of Flower 28
  - 2.1.1 The Color 28
  - 2.1.2 Flower's Shapes 30

- 2.2 Conventional Herbal Cutting Flower Material ... 31
- 2.3 Conventional Woody Cutting Flower Material ... 31
- 2.4 Conventional Equipment for Flower Arrangement ... 39
  - 2.4.1 Conventional Equipment for Flower Arrangement ... 39
  - 2.4.2 Conventional Tools and Their Accessories ... 40

Chapter 3 Flower Arrangement for Etiquette

- 3.1 Wedding Flower Arrangement ... 46
  - 3.1.1 Decorative Flower for Bride ... 46
  - 3.1.2 Floats/Festooned Vehicle ... 48
  - 3.1.3 Brooch ... 51
  - 3.1.4 Strewing Flower ... 52
- 3.2 Celebration Flower ... 52
  - 3.2.1 Features ... 52
  - 3.2.2 Celebration Flower Basket ... 54
- 3.3 Flower Arrangement for Funeral Affairs ... 55
  - 3.3.1 Wreath ... 56
  - 3.3.2 Bouquets in Basket ... 56
  - 3.3.3 Flower Arrangement ... 57
- 3.4 Flowers for Conferences and Banquets ... 57
  - 3.4.1 Features ... 57
  - 3.4.2 Common Styles ... 58
- 3.5 Birthday Flower Arrangement ... 59
  - 3.5.1 Features ... 59
  - 3.5.2 Common Styles ... 61
- 3.6 Flowers for Visiting Patients ... 62
  - 3.6.1 Features ... 62
  - 3.6.2 Common Styles ... 62
- 3.7 Flower Arrangement on Other Occasions ... 62

Chapter 4 Propositions of Artistic Flower Arrangement and the Connotations of Flower Culture

4.1 Propositions of Artistic Flower Arrangement ················· 66
- 4.1.1 Characteristics of Proposition ················· 66
- 4.1.2 Requirement for Proposition ················· 66
- 4.1.3 Basic Methods of Proposition ················· 67
- 4.1.4 Specific Methods of Proposition ················· 67
- 4.1.5 Ways to Improve the Accomplishments of Proposition ········· 68

4.2 Appreciation of Artistic Flower Arrangement and Flower Culture 69
- 4.2.1 Flower Appreciation and Flower Culture ················· 70
- 4.2.2 Gist of Flower Appreciation ················· 71
- 4.2.3 Manifestation of Cultural Connotations of Flowers in Flower Arrangement Works ················· 72

Chapter 5 Fundamentals and Commonly Used Techniques of Flower Arrangement

5.1 General Introduction to Skills of Flower Arrangement ················· 82
- 5.1.1 Processing of Floral Materials ················· 82
- 5.1.2 The Structure of Floriculture ················· 83

5.2 Commonly Used Techniques of Expression ················· 85
- 5.2.1 Three-dimensional Composition Technique ················· 85
- 5.2.2 Highlighting-One-Branch Technique ················· 87
- 5.2.3 Oblique Angle Echoing Technique ················· 88
- 5.2.4 Grouping Technique ················· 89

5.3 Basic Skills of Flower Arrangement ················· 91
- 5.3.1 Preparation ················· 91
- 5.3.2 Trimming of Floral Materials ················· 91
- 5.3.3 Fixing of Floral Materials ················· 94
- 5.3.4 Finishing Works ················· 95

Chapter 6 Preservation and Packaging of Flowers Meterials

6.1 Preservation of Flower Materials ················· 98
- 6.1.1 Causes of Flower Materials Wilting ················· 98

	6.1.2	Principles of Preserving Flower Materials	101
	6.1.3	Methods of Preserving Flower Materials	103
	6.1.4	Preservative Solutions for Flowers	106

6.2 Flower Packaging 112

	6.2.1	Flower Packaging and Consumer Psychology	112
	6.2.2	Modern Packaging Materials	116

Color Illustration 127

第1章 插花艺术概述

Chapter 1　A General Introduction to Flower Arrangement Art

插花作为人类的一种文化活动，由来已久。无论在东、西方，均有约2000年的历史。人们以剪切植物为素材，经过艺术加工，赋予这些素材文化内涵，形成了一门独特的艺术——插花艺术，在中国称为插花，在日本称为花道，在欧美国家称为花艺。

插花艺术与书法、绘画等平面艺术不同，它是一门立体艺术，但又与雕塑、建筑等立体艺术不同，它是一门有生命的立体艺术。因此，作者不但要遵循美学法则，进行艺术创造；要吸收文学精髓，赋予精神内涵；还要掌握植物的生理、生态特征，合理加工，予以恰当表现，并延长观赏期。从这个意义上说，插花艺术综合的是艺术和技术。

插花艺术既是一门学科，也是一个创意产业。对插花者而言，既要掌握构图、色彩、植物、文学等知识，同时又要对材料、光学、力学等知识有所了解，结合自身的悟性，方能创作出好的作品。考虑到文明社会的建设和小康生活的需求，社会将需要大批插花人才，尤其是高级人才。

1.1 插花艺术的定义

插花艺术是以剪切花木为素材，经过艺术加工，并赋予文化内涵的造型艺术。它包括以下三方面的内涵：

首先，插花艺术的创作素材既不同于绘画、雕塑、建筑等艺术，也不同于以带根植物为素材的盆景、造园、组合盆栽等艺术。它是以剪取下来的植物根、茎、花、叶、果为素材，是有生命的艺术品。

Flower arrangement, as a kind of cultural activity of human beings, has a long history. It has a history of two thousand years in both the east and the west. In flower arrangement, people cut plant materials through artistic processing, and give them cultural connotation, which has been developed into a unique art called flower arrangement (Chahua) in China, ikebana in Japan, and floriculture in European countries and USA.

Flower arrangement art is a three-dimension art, which is different from graphic arts (like calligraphy and painting). Unlike sculpture and architecture, it is a living three-dimension art. Therefore, artists must not only follow the laws of aesthetics to create art works, but also absorb the essence of literature and give spiritual connotations, grasp the plant physiological and ecological characteristics, rational processing and proper performance so as to extend the viewing period. In this sense, flower arrangement art integrates art with techniques.

Flower arrangement art is both a discipline and a creative industry. The flower arranger must not only master the knowledge of composition, color, plant, literature, and the like, but also gain some understanding of materials, optics, mechanics, and the like. Only by combing his or her own understanding can the flower arranger create good works. With the construction of a civilized society and demand for well-off life, a large number of flower arranging talents will be needed, especially high-level talents.

1.1 Definition of Flower Arrangement Art

Flower arrangement art is a formative art of using flowers as the material through artistic cutting or processing and giving it cultural connotations. It includes the following three aspects:

First, the raw materials of flower arrangement are different from that of painting, sculpture, and architecture, and also from that of Penjing, gardening and assembled Penjing which use plants with root as the material, for it uses the roots, stems, flowers, leaves, and fruits cut from plants as the material, and creates animated artistic works.

其次，插花艺术并非买一把花，随意往瓶里一插即可。而是以花草为创作素材，遵循美学原理和色彩理论，对原材料进行加工、修剪和整理，排列组合，以崭新的植物艺术形态出现，满足人们不断提高的审美情趣，获得高品位的精神享受。

最后，插花艺术的特性不仅具形态美和色彩美，还要求具有内在的意境美。意境美是作者创作心意的写照，是主观和客观的统一，它赋予插花作品更深的感染力，让人们过目难忘。中国插花艺术就特别重视意境美的创造。

Second, flower arrangement art does not mean buying a bunch of flowers and randomly inserting them into a vase. It is based on flowers and plants, following the principle of aesthetics and color theory. Raw materials should be processed, cut, sorted, permuted, combined and displayed with a brand-new artistic form so as to satisfy people's increasing aesthetic tastes and help them to get high quality spiritual enjoyment.

Moreover, the characteristics of flower arrangement art do not only include the beauty of form and color, but also involve the inner beauty of artistic conception. The beauty of artistic conception is the portrayal of the artist, the unity of subjectivity and objectivity. It gives the flower arrangement work stronger appeal, making it unforgettable. Such as, Chinese flower arrangement art pays special attention to creating the beauty of artistic conception.

1.2 插花艺术的历史

1.2.1 中国插花历史简介

中国插花起源于南北朝，盛行于唐宋，普及于明清，衰微于清末，复兴于改革开放的20世纪80年代初。

（1）南北朝——中国插花的形成时期

中国插花至今已有1500多年历史，中国是东方插花的起源国。中国历史上有关插花的记载，最早见于5世纪的《南史》。《南史》晋安王子懋传中："有献莲华供佛者，众生以铜罂盛水，渍其茎，欲华不萎"。铜罂，原是一种口小腹大的盛器，可以盛酒储水，也可以插花供佛。子懋，齐武帝第七子，七岁时，母亲淑媛病危，请僧侣为母祈祷。当时有人以莲花献佛，众僧以铜罂盛水，以维持花色新鲜。子懋流涕礼佛，并称："若母能因此病愈，此花当于斋毕之前，仍能

1.2 History of Art of Flower Arrangement Art

1.2.1 A Brief History of Chinese Flower Arrangement

Chinese flower arrangement originated from the Northern and Southern Dynasties. It was prevalent in the Tang and Song Dynasties, popular in the Ming and Qing Dynasties, and declined in the late Qing Dynasty. It was revived in the period of the Reform and Opening-up in the early 1980s.

(1) The Northern and Southern Dynasties—the formation period

Historically, Chinese flower arrangement art has a history of more than 1500 years. China was the original country of flower arrangement in the East. There is a recorded history in China concerning flower arrangement: it appeared in the 5th century AD in *The History of the Southern Dynasties*. There is one section of Jin'an Prince Zi Mao's biography that says "People offer lotus to Buddha, all beings fill copper vase with water, and cut the stem of lotus to make it stay fresh." Here copper vase refers to a container with small opening and big stomach, in which water or wine can be stored, or arrange a flower for Buddha. Zi Mao was the seventh son of Emperor Qiwu. At the age of seven, his mother Shu Yuan was dying, he invited monks to pray for his mother. When one monk offered lotus to Buddha, others filled copper vase with water

维持新鲜不萎。"当时的插花，只是把花插在花瓶里，略加调理而已，尚谈不上艺术插花。

随着历史的发展，插花在佛教供花的基础上，在民间流传风气渐盛。人们在户外赏花之余，把花草摘回插在瓶、盆中观赏，形式上也逐步摆脱供花格式，纯作观赏之物。

（2）隋唐五代——中国插花艺术的黄金时代

插花的种类及形式发展很快。除了寺庙的祭坛供花外，插花在宫廷生活，尤其是赏宴中，非常流行，且种类繁多，有瓶花、篮花、缸花、挂花、竹筒花等。此时民间赏花也极兴盛，在唐朝中期，定二月十五日为"花期"，即百花诞生的日子。当天，男女老少倾城而出，视赏花为一大福事。正如诗人杜牧在《杏园》诗中描述："夜来微雨洗芳尘，公子骅骝步贴匀，莫怪杏园憔悴去，满城多少插花人"。

这一时期，人们对花木已普遍赋以人格象征，这是插花上的一大突破。由此，插花配材更为审慎，既要知道它们的自然习性，还要了解它们的人文属性。插花也已被视为一门学问而得以研究。一些插花专著先后问世，如唐代欧阳詹的《春盘赋》和唐代罗虬的《花九锡》等。

to keep the color fresh. Zi Mao cried to pay respect to Buddha with runny nose and said: "If my mother recovers from illness, the flower will still be fresh before the fast ends." Flower arrangement at that time was, however, just putting the flowers in a vase with proper adjustment, which is not real artistic flower arrangement.

With the development of history, based on flower arranging for Buddhism, flower arrangement became popular among the folks. While viewing flowers outdoor, people began to pick flowers, take them home, and then insert them into vases or basins to appreciate, and gradually people arranged flowers for pure appreciation instead of worship.

(2) The Sui, Tang and Five Dynasties—the colden age

Different types and forms of flower arrangements developed rapidly. Besides worship ceremonies, arranged flowers were also used in court life, especially popular in the feast with a great variety, such as flowers arranged in bottles, baskets, vases, jars, bamboos, and so on. At this time, viewing flowers among folks were also prevailing. In the mid-Tang Dynasty, February 15th was named as "Flower Season", in which all kinds of flowers were blooming. On the same day, people of all ages and both genders came out to appreciate flowers as a great blessing. As the poet, Du Mu, described in the poem *Apricot Orchard* (in Chinese "Xing Yuan"): A light rain washed dusts overnight; a child riding a red horse was enjoying the scenery of the apricot orchard with leisure and relaxation; do not blame it for its emaciation; owing to the beauty of apricot orchard, many people viewing the flowers took apricot flowers away for their own momentary pleasure.

During this period, people had endowed flowers and trees with a symbol of personality, which was a breakthrough in the history of Flower Arrangement Art. Thus, materials for flower arrangement were more carefully chosen. The arrangers needed to know both their natural habits and cultural properties. Flower arrangement had been considered as a specialized knowledge to carry out further researches. Some monographs about flower arrangement were published successively, such as Ouyang Zhan's *Ode to Spring Plates* and Luo Qiu's *Nine Things for Flowers* in the Tang Dynasty.

用于奉佛的插花,花材较少,每瓶花材只限一种。尤其是莲花,枝叶不多,常以简洁明快的三支枝干为主要架构,最长者为花,其余两支为叶。一般宫廷或家庭插花,花材稍多,以牡丹、兰、梅、莲、桃、杏等为主要花材,配以松、柳、槐等叶材。插花形式除以三角形排列外,也有向四面八方多角度开展的,显得隆重而豪华。插花结构比例,花枝与花器约为8:5或5:8,合乎黄金比例。

五代是中国插花艺术的成熟期。虽然当时战势纷乱、民不聊生,插花的普及程度不如隋唐,但此时一批文人墨客介入插花,他们以花明志、以花浇愁,并在花器、花材、花形方面不拘一格,自由发挥,恰好使插花艺术达到自由奔放的成熟境地,"自由花"的时代由此诞生。当时的"自由花"最大的特点是不讲排场,不求奢华,讲究生活情趣和花草兴趣。花器以朴质为主,花型简洁利落,花色淡雅,处处表现出文人的怡性风雅。由于很多文人墨客本身就是画家、诗人、文学家,他们精通美学原理,并在插花活动中注入美学法则,从而推进了插花艺术的发展。

Fewer materials were used in the flower arrangement for Buddhist worships, and only one type of flowers was arranged in each vase. *Nelumbo nucifera* (Lotuses) in particular, with a few leaves, were often arranged with three branches as main framework in a simple style. The longest one was the flower, and the other two were leaves. Floral materials used in court or family flower arrangement were relatively various, which mainly included *Paeonia suffruticosa* (peony), *Cymbidium* ssp. (orchid), *Armeniaca mume* (plum blossom), *Nelumbo nucifera*, *Amygdalus persica* (peach blossom), *Armeniaca vulgaris* (apricot blossom), and so on, and were decorated with leaves of *Pinus* spp. (pine), *Salix babylonica* (willow), *Sophora japonica* (sophora japonica), etc. The forms included triangle arrangement, as well as multi-angle arrangement, which appeared grand and luxurious. The proper structural proportion of floral materials to the vessel was 8:5 or 5:8, which is corresponding with the golden ratio.

The Five Dynasties was the mature stage of flower arrangement art. Although it was in chaos because of wars and people had a hard time, and flower arrangement were less popular than that in the Sui and Tang Dynasties, a group of men of letters were involved in it. They showed their ideals and relieved their sorrows through flowers, and followed no set form in the selection of vessels, floral materials and flower shapes. It made the art evolve into a mature period with freedom and unrestraint, and helped the coming of the "Free Flower" Era. The biggest feature of "Free Flower" is seeking the delight of life and the interests in flower instead of ostentation and luxury. The vessels were usually plain, the flower shapes were simple and neat, and the colors were light and elegant, which demonstrated scholars' refined and elegant taste of life. Since many scholars themselves were painters, poets, and writers, who had a good command of aesthetic principles, they integrated aesthetic laws into flower arranging activities and promoted the development of flower arrangement art.

此外，五代在中国插花史上最有影响的是"锦洞天"和"占景盘"。

"锦洞天"是南唐后主李煜独创。他于每年盛春在宫殿举行盛大插花展，这是我国最早的插花展。《清异录》记载："李后主每春盛时，梁栋窗壁、柱拱阶砌，并作隔筒，密插杂花，榜曰：'锦洞天'。"可见，早在1000多年前，中国不但已有插花展览，还有空间的花艺布置。

"占景盘"是插花固定工具的一大发明。它是由五代郭江洲发明的，他在盘底铸满许多低矮铜管，使花枝插在其中不致倒塌。"占"是站的别字，"占景"意为竖立的花草景物。占景盘的发明，使插花造型或写景表现更为容易，它是中国插花艺术上的一大进步，是花插（剑山）的雏形。

（3）宋代——中国插花艺术的鼎盛期

五代战乱平寂后，民心思定，人们将赏花作为一年中的"赏心乐事"，尤到"花朝"，相率赏花，遍成习俗。

宋代把插花与焚香、点茶、挂画合为生活四艺，视为每人从小就应具备的修养。插花展也逐渐兴盛，有些地方还办起了有数万枝花的"万花会"。

此时，宫廷插花分日常插花和节日插花。日常插花以装饰为主，包括宫室摆设与宴会插花。节日插花，尤其是"花朝"，插花隆重豪华，气魄非凡。连平时供赏的古董瓶盆，也用来插上名贵花材，作为插花观赏。宫廷插花风格，以形体硕大、色彩壮丽、枝叶繁茂、结构严谨的院体花为特色。盘花粗看与西洋插花相似，细看不然，其花色为不等边三角形布局，花枝也

In addition, "Jin Dong Tian" and "Zhan Jing Pan" of the Five Dynasties are the most influential in the history of Chinese Flower Arrangement.

"Jin Dong Tian" comes from Li Yu, the emperor of the Southern Tang Dynasty, who held grand flower arrangement shows in mid-spring annually in the palace, which was the earliest flower arrangement show in Chinese history. *Qing Yi Lu* recorded: "In every mid-spring, Emperor Li would decorate almost every window, wall, column and step with flowers, and named the scene as "Jin Dong Tian" (meaning "a magnificent heaven of colorful flowers"). Therefore, as early as 1000 years ago, there had been both flower arrangement shows and spacial floral arrangement in China.

"Zhan Jing Pan" was a great invention of fixing tools for flower arrangement. It was invented by Guo Jiangzhou in the Five Dynasties. He cast many low copper pipes on the bottom of a plate, in which flowers could be arranged straightly. "Zhan" means standing straight, and "Zhan Jing" means spectacular erect plants. This invention made flower design and arrangement or display of flower scenery easier, marking a great improvement in China flower arrangement art, and it was also the prototype of flower receptacle.

(3) The Song Dynasty—the prosperous period

After the end of the war in the Five Dynasties, people lived in peace. They viewed the appreciation of flowers as the most enjoyable event in the year. Especially on the "Flower Day", most people went out to enjoy the flowers, and so it became a traditional custom.

In the Song Dynasty, flower arrangement, burning incense, boiling tea, and hanging pictures were seen as the four skills in life which everyone had to possess from an early age. Flower arrangement shows gradually became prevailing. In some places, people held flower shows with hundreds of thousands of flowers.

At that time, flower arrangement in palace included daily flower arrangement and festive flower arrangement. The first one focused on decoration, including palace decoration and banquet arrangement. The other one focused

非直线状放射，而是成曲折逶迤布置。花与枝间或疏或密，相得益彰。花叶之间，即使密插，也是朵朵舒立，不显挤乱。而花材的姿态、色彩、大小、向背绝忌雷同，充分显示作品的韵律感和生命感。

宋代文人插花讲究清雅，往往寥寥数枝，也足以陶咏花间，畅诉幽情，杨万里诗云："胆样银瓶玉样梅，北枝折得全未开，为怜落寞空山里，唤入诗人几案来。"

寺院及民间插花此时也更为普及，佛寺除供花外，禅师在禅房、经桌或户外，也用插花布置。作品用材不多，寥寥数枝，至富禅意。民间插花在"花朝"和节庆最为热闹，常年在不同节庆插不同花，流行之风气遍及主要城市。

宋代的瓶花，出现了理念花的新风格，即在花瓶插花中，寄寓解说教义、阐述教理、影射人格、诉说哲理等，以理为表，以意为里，内容重于形式，称之为"理念花"。"理念花"以"清"为精神之所在，以"疏"为意念之依归。作品注重线条肌理，脉络分明、神圣雅洁、调理有序。

on grand, luxurious and magnificent arrangement, especially on the "Flower Day". Even the antique bottles and basins that were usually used for appreciation were now used as vessel for those rare and precious floral materials. The style of palace flower arrangement features court flowers which had large body, glorious colors, flourishing branches and compact structure. Plate flower arrangement looked similar to western flower arrangement with a rough glance. But with a closer look, the colors of the flowers in plate flower arrangement formed a scalene triangle, and the sprays were also arranged windingly rather than linearly. Space between flowers and branches were either sparse or dense, complementing each other. Even though flowers and leaves were arranged densely, it did not look crowded with standing blossoming flowers. The poses, colors, sizes and the facing directions were always different from one another, which fully displayed the work's rhythm and vitality.

Scholars of the Song Dynasty were particular about elegance in flower arrangement. Often, only several simple branches and flowers could be enough for them to recite poems and express their exquisite feelings freely, just like what Yang Wanli's poem conveyed: "Jade-colored plum blossoms standing in a silver pear-shaped bottle; I cut them down from the tree while they are not fully bloomed; it is so cold and lonely in the mountain, and it's better to gather several poet friends to chat with around the table."

Flower arrangement in temples and among folks had become more and more popular. In temples, besides the flowers for worshiping the Buddha, Buddhists decorated their rooms, desks or the outdoor places with flowers. Simple and plain, their flower arrangements did not contain many floral materials, but were full of Buddhist mood. Folk flower arrangement was most prevailing during the "Flower Day" and festivals. Different flower arrangements were made for different festivals. Its popularity permeated all the major cities.

Vase flower in the Song Dynasty began to show the new style of conceptual flower, which tried to explicate

宋代人们对花品、花性的研究更为深入，花的拟人化与分类品评之风甚盛。如曾慥以兰花等十花为十友："兰为芳友，梅为清友，瑞香为殊友，莲为净友，栀子为禅友，蜡梅为奇友，菊为佳友，桂为仙友，海棠为名友，荼䕷为韵友。"黄庭坚以梅花等十花为十客："梅花索笑客，桃花销恨客，杏花依云客，水仙凌波客，芍药殿香客，莲花禅社客，桂花招隐客，菊花东篱客，兰花幽谷客，荼䕷清叙客。"其他诸如对花的品种与品第研究、插花器具研究、瓶花保养研究等，在宋代都有较大发展。

doctrines, allude to personality, and convey philosophy via the moral of the flower arrangement. "concept flower" expressed natural science on the surface, but conception inside, putting more emphasis on the content than the form. It took elegance as the spirit and alienation as the concept. Concept flower arrangement works focused on texture lines, clear structure, holly elegance and good order.

People in the Song Dynasty made further research on the characteristics and natures of flowers. Personification of flowers and classified judging were particularly prevailing. For example, Zeng Zao took *Cymbidium* ssp. and other nine kinds of flowers as friends: "*Cymbidium* ssp. — the friend of fragrance; *Armeniaca mume* — the friend of elegance; *Daphne odora* (daphne) — the friend of difference; *Nelumbo nucifera* — the friend of purity; *Gardenia jasminoides* (gardenia) — the friend of Buddhism; *Chimonanthus praecox* (wintersweet) — the friend of particularity; *Dendranthema morifolium* (chrysanthemum) — the friend of excellence; *Osmanthus fragrans* (osmanthus flower) — the friend of immortality; *Malus spectabilis* (malus spectabilis) — the friend of good reputation; *Rosa rubus* (roseleaf raspberry) — the friend of rhythm ." Huang Tingjian regarded ten kinds of flowers as guests: "*Armeniaca mume* could make people smile; *Amygdalus persica* could make people forget unhappiness; *Armeniaca vulgaris* could make people have more friends; *Narcissus tazetta* (narcissus/ daffodil) could make people relax like walking over ripples; *Paeonia lactiflora* (peony blossoms) could make people smell sweet in palaces; *Nelumbo nucifera* often could be guests in meditation rooms; *Osmanthus fragrans* could attract hermits; *Dendranthema morifolium* could be unworldly guests; *Cymbidium* ssp. could be guests in deep and secluded valleys/the hidden glens; *Rosa rubus* could be men of noble character." Moreover, there were greater developments in researches on the varieties of flowers, flower vessels and maintenance of vase flowers in the Song Dynasty.

元代，中国花艺文化遭到空前浩劫。人们无暇种花，无心赏花，唯能勉强维持插花活动的只有隐居山林、位高富贵的士大夫，他们在逆境中自种花草，折花插花，孤芳自赏，以花浇愁。特别在插花中，借用花语花性，以花示意、以花遣兴，由此产生了"心象花"。

（4）明代——中国插花艺术的复兴期

虽然明初朱元璋限制宫廷造园种花，宫廷插花已无前代风光，只在喜庆节日进行。而此时在农业发展的推动下，民间的花卉栽培得到了空前的发展，民间插花也进而普及。明初有建花神庙，在花朝祭花神的风俗，热闹非凡。

这个时期，文人思想解放，复古气氛弥漫，养性摄身之学流行，研究插花风气也随之昌盛，插花专著也不断问世。如金润的《瓶花谱》、高濂的《瓶花三说》《草花谱》、屠隆的《考·余事》、张谦德的《瓶花谱》、袁宏道的《瓶史》、文震亨的《长物志》、何仙郎的《花案》、屠本的《瓶史月表》、陈续儒的《岩栖幽事》等。

In the Yuan Dynasty, Chinese floriculture suffered from an unprecedented catastrophe. People were too busy with their own lives to plant or appreciate flowers. Only the officials who lived in seclusion and had superior status with wealth could sustain the activity of flower arrangement. They grew plants and flowers on their own, then cut flowers and arranged them for appreciation to dispel unhappiness. They used flower languages and flower characters especially in flower arrangement to express their inner feelings and interests, which resulted in a style named "mood flower" ("Xin Xiang Hua" in Chinese).

(4) The Ming Dynasty—the revival period

Although at the beginning of the Ming Dynasty, Emperor Zhu Yuanzhang restricted the building of garden and planting flowers in the palace, making palace flower arrangement not as popular as before, and flower arrangements could be seen only in festivals and celebrations. But during that time, with the promotion of agricultural development, floriculture had been developed in folks at an unprecedented rate, and then flower arrangement among the people was popularized. In the early Ming Dynasty, Flower God Temple was built, and people offered sacrifices to the Flower God on the Flower Day.

During this period, with the ideological emancipation of the scholars, there appeared a vintage atmosphere, and the knowledge of temperament moulding began to be popular. Thereupon, the study of flower arrangement also became prosperous, and the monographs on flower arrangement were published one after another. For example, there were Jin Run's *Anthography of Vase Flowers*, Gao Lian's *Three Comments on Vase Flowers* and *Anthography of Grass Flower*, Tu Long's *Things about Living in Seclusion*, Zhang Qiande's *Anthography of Vase Flowers*, Yuan Hongdao's *History of Vase Flower*, Wen Zhenheng's *Superfluous Things*, He Xianlang's *Archive of Flowers*, Tu Benjun's *History of Vases in Month Table*, and Chen Xuru's *Things about Inhabiting in Rocks*.

其中,《瓶史》一书于1600年出版后,不仅在国内轰动一时,促成文人插花风气流行,并于1696年,被译为日文出版,对日本"袁宏道流""宏道流"插花艺术流派的出现影响至深。

明代文人插花,既讲外观,更讲内涵,其创作意义常包含作者的理想愿望,虽与宋代理念花近似,但花形更加简洁而注重花德,被称为"新理念花"。"新理念花"为了追求古典、完整的插花形式,还注重通过自然美来表现人类社会的规律美、人格美,使人类社会更加和谐。花中蕴含作者对社会秩序的希望与处世哲学,不仅是个人的,还是社会的、理论的、善良的、出世的,意义积极。

文人插花构图疏松,给人以清新俊逸之感。为追求雅逸之趣,花材经常以带散花或散叶的枝条为主,以追求线条美和虚灵美。

明代插花,以瓶花为主。当时花瓶有两大类型,称为"金屋"与"精舍"。"金屋"指雍容华贵的铜瓶和景泰蓝瓶等,用于殿堂、厅堂、庙堂等隆重场合。"精舍"指高雅朴素的陶瓷等小花器,用于书房、斋房等清雅之所。

(5)清代——中国插花艺术的衰微期

清初插花仍较兴盛,晚清慈禧时期也流行插花,鸦片战争以后,随着政治经济的衰颓,插花艺术日显沉寂。

Of them, *History of Vases* was published in 1600, which not only made a great flutter at home, and promoted the popularity of flower arrangement among scholars, but also was translated into Japanese in 1696 by Japanese, influencing Japanese flower arrangement so deeply that flower arrangement schools of Yuan Hongdao and Hongdao came into being in Japan.

The scholar flower arrangement of the Ming Dynasty paid attention to both appearance and connotation. Its creation significance often contained the arranger's ideals and aspirations. Although it was similar to the "concept flower" of the Song Dynasty, its form was more concise and focused on flower virtue, so it was called "new concept flower". In order to pursue classical and complete flower arranging form, "new concept flower" also paid attention to expressing human society's beauty of law and personality with the symbol of natural beauty to make mankind more harmonious. The flowers implied the arranger's hope for a sound public order and philosophy of life, for both individuals and society, which was theoretical, kind, worldly, and positive.

Scholastic flower arrangement had a loose composition, with a fresh and unsophisticated sense. In pursuit of elegance, branches with scattered flowers or leaves were usually chosen as floral materials to create linear beauty and ethereal beauty.

Vase flower arrangement played a dominant role in the Ming Dynasty. There were two major kinds of vases: one was called "golden room" and the other "vihara". The former referred to gorgeous brass vases and cloisonne vases used in ceremonious places such as palace, auditoriums, and temples. The later referred to the grace and simple small vessels made of porcelain or other materials used in elegant places such as study rooms and fast rooms.

(5) The Qing Dynasty—the declining period

Flower arrangement was still thriving in the early and late Qing Dynasty. After the opium war, along with the political and economic decline, flower arrangement began to decay.

在清初插花兴盛期，宫廷插花风格继承了明代"理念花"，器具较为华丽，花材色彩偏爱红、黄等明丽颜色。民间插花盛于宫廷，在有些节日，民间与宗教寺院合并举行插花展览和比赛，景象热闹。

　　受盆景艺术发展的影响，此时写景花更为流行，既有不带根的写景插花，也有带根的组合盆栽。同时，谐音造型花和蔬果插花开始兴起。

　　清代插花在艺术创作方面虽属停滞时期，但品赏花的方法和插花理论研究则不逊于前代。如将花人格化，进而神格化，即把每一种花与历代名人相配，作为花神。每年12个月，以12种花配对12个花神，按月颂扬祭拜。

　　清代插花专著有李渔的《闲情偶寄》、张潮的《幽梦影》、沈复的《闲情记趣》等。

　　清代末期，由于战乱频繁，民生困苦，插花艺术也日渐衰落。

　　（6）新中国——插花艺术的复苏发展时期

　　20世纪80年代初，随着改革开放的深入，中国经济文化迅速崛起，中国插花艺术也迎来了又一个春天。北京、上海、广州等几个大城市，率先成立插花组织，研究插花艺术，举办插花展览，创办插花培训班。政府和社会各方积极支持，群众积极响应，使插花艺术这一优秀的传统文化迅速复苏并发展。如今，除了有千百万插花爱好者参与这一文化活动之外，还形成了一只专业的插花产业军，经过培训和考核鉴定，由政府授予职业资格证书。同时，在北京、上海等一批高等院校还设置插花花艺大专班，

　　In the early Qing Dynasty when flower arrangement was prosperous, the style of palace flower arrangement was inherited from the "concept flower" of the Ming Dynasty. Flower utensils were gorgeous; colors were bright like preferential red and yellow. Flower arrangement among the folks was more flourishing than that in the palace. In some festivals, folks and religious temples co-held flower arrangement shows and competitions lively.

　　With the influence of bonsai art development, scenery-portrait flower arrangement became more popular. There were both rootless scenery-portrait flower arrangement and combination of potted plants with roots. Meanwhile, homophonic modeling flowers and flower arrangement with vegetable and fruit began to spring up.

　　In art creation aspect, flower arrangement in the Qing Dynasty entered a period of stagnation, but the ways of flower appreciation and arrangement theories were not inferior to the previous generations. For example, flowers were personified and then deified, matching each kind of flower with famous people in the history as Flower Gods. There are twelve months in a year, so there were twelve kinds of flowers and twelve Flower Gods for people to worship monthly.

　　In the Qing Dynasty, there were many flower arranging monographs, such as Li Yu's *Occasional Enjoyment in a Free Mood*, Zhang Chao's *Dream Shadows*, Shen Fu's *The Little Pleasures of Life*.

　　In the late Qing Dynasty, due to frequent wars, people lived in miserable conditions. Flower arrangement art began to decline.

　　(6) New China—the recovery and developing period

　　In the early 1980s, with the deepening of the reforms and openings-up, China's economy and culture grew rapidly, and Chinese flower arrangement art was also recovering. Several major cities such as Beijing, Shanghai, and Guangzhou firstly established flower arrangement organizations for studying flower arrangement art, holding flower arrangement exhibitions, and starting flower arrangement training. With the support from the

培养高级插花人才。另外，各地对外插花交流也蓬勃发展。插花已成为一种创意产业，受到政府和社会的高度重视。

1.2.2 日本插花历史简介

日本插花起源于中国。6～7世纪，由日本遣唐使小野妹子把插花随同中国的佛教文化传入日本。小野妹子完成使节任务后，皈依佛门，居住在京都由圣德太子创建的六角堂寺院的池坊，潜心佛道修养，并吸收中国佛教供花的精髓，结合日本习俗制定了祭祀插花时花材配置的种种规矩，创建了第一个插花流派——池坊流。因此，日本插花始于池坊，池坊插花的历史一定程度上代表了日本插花的历史。小野妹子是日本插花的始祖。

日本插花在相当长的时间内，仅限于庙宇供奉之用，具有浓厚的宗教色彩。到了9世纪，插花才逐渐摆脱宗教的束缚，从庙宇传到宫廷和将军贵族官邸。后随中国瓷器大量输入日本，插花又变成装饰花瓶的观赏花。14世纪，足利义满将军成立了"七夕观花会"，从此，插花成为举行典礼仪式必备的装饰品之一。

government, different social sectors and the masses, flower arrangement art, an excellent traditional culture, recovered rapidly and developed. Today, besides millions of flower arrangement enthusiasts participating in this cultural activity, a group of professional flower arrangers has been developed. After being trained and assessed, they have been awarded the professional qualification certificate by the government. Meanwhile, a number of colleges and universities in Beijing, Shanghai and other cities also offer junior college courses of flower arrangement to train senior flower arrangement talents. In addition, international flower arrangement exchanges are also flourishing. Flower arrangement has become a creative industry highly regarded by the government and the society.

1.2.2 A Brief History of Japanese Flower Arrangement

Japanese flower arrangement originated from China. In the 6th~7th century, Japanese ambassador Onono Imoko, who was sent to China in the Tang Dynasty, introduced flower arrangement along with Chinese Buddhist culture to Japan. After he finished his diplomatic mission, he was converted to Buddhism, living in Ikenobo of Temple founded by Prince Shotoku in Kyoto, and devoted himelf to Buddhism and absorbed the essence of flower offering in Chinese Buddhism. In combination with Japanese customs, he made rules of the floral materials for arrangement in worship ceremonies and created the first flower arrangement school — Ikenbo. Therefore, Japanese flower arrangement started in Ikenobo, and the history of flower arrangement in Ikenobo represents Japanese flower arrangement history to some extent. Onono Imoko is the earliest ancestor of Japanese flower arrangement.

For quite a long time, Japanese flower arrangement was limited to the usage for offering to Buddha with strong religious connotation. By the 9th century, flower arrangement was gradually free from the bondage of religion, being spread from temples to the palace and the residences of the

15世纪，日本室町时代完成了日本插花的重要样式——立华。立华是池坊流的代表花型，通常由7~9个主枝构成，构图严谨，意念抽象，着力表现大自然的景观之美。

17世纪末，我国明代文学家袁宏道的插花专著《瓶史》传入日本，日本插花界大为推崇，创立了"袁宏道流""宏道流"等插花流派，使《瓶史》在日本得到发扬光大。

17~18世纪以后，为了普及插花，在立华的基础上又相继出现了"生花""投入花"和"自由花"等形式。

19世纪，出现了盛花形式，即用剑山把花插在浅盆观赏。

第二次世界大战后，随着日本经济的复苏，插花之风更盛。流派众多，号称有3000流派，其中最有代表性的是三大流派——"池坊流""小原流"和"草月流"。

日本花道在中国插花基础上，以儒家的伦理学为哲学基础，禅宗佛教为指导，通常以三主枝分别代表天、地、人，使人和自然成为和谐的整体。三主枝的高低俯仰、曲折变化，形成了不同的花型。

generals and nobles. Then with Chinese porcelain being imported into Japan, flower arrangement evolved into ornamental flowers in vases. In the 14th century, General Ashikaga Yoshimitsu created the "Tanabata Flower Show". From then on, flower arrangement became one of the necessary ceremonial ornaments.

In the 15th century, the important flower arranging style — Tachibana was completed in Japan Muromachi Period. It was the typical style of ikeobo, which consists of 7 to 9 main branches with a rigorous composition and abstract conception, focusing on the expression of the beauty of natural landscape.

In the late 17th century, Chinese well-known litterateur Yuan Hongdao's monograph *History of Vase Flower* of the Ming Dynasty was introduced to Japan. Japanese ikebana world greatly respected this monograph, and created "Yuan Hongdao School" "Hongdao School" and other flower arrangement schools. After that, *History of Vase Flower* was widely spread in Japan.

After the 17th~18th century, in order to popularize flower arrangement, on the basis of Tachibana, there appeared some other styles of flower arrangement one after another such as "Ikebana" "nageire" "free flower", and so on.

In the 19th century, moribana style appeared, which referred to inserting flowers into shallow pots with Tsurugi-san (a styliform frame) for appreciation.

After World War II, with the recovery of Japan's economy, the custom of flower arrangement became more prevalent. There were about 3000 schools of flower arrangement, of which there were three representative schools — "ikenobo" "Ohara" and "Sogetsu".

Japan ikebana was established on the basis of Chinese flower arrangement with Confusion philosophy, guided by Zen Buddhism. Generally, the three branches represented heaven, earth, and human being, which made people and nature become one harmonious whole. The height, facing direction and inflection of the three main branches form different three axial shoots are high and low, pitching and tortuous, forming different patterns.

随着战后大量鲜花涌入日本，新的花材观、造型观也逐步形成，出现了前卫插花，即以造型为主的现代插花。

1.2.3 西方插花历史简介

西方插花起源于古埃及和古希腊。埃及人很早就有将睡莲花插在瓶或碗里，作为装饰品、礼品或丧葬品的习俗。以后随着文化的传播，插花先后传到比利时、荷兰、英国等。

插花早期在欧洲流传，多作为宗教用花。西方人认为花可以驱除巫术和闪电，常用橄榄枝和月桂枝作为花环戴在头上或脖子上作为护身符，或者挂在门、墙上，防邪魔进入。

14～16世纪的欧洲文艺复兴运动时期，插花摆脱了宗教的束缚，得到了迅速发展。插花风格受西方艺术中几何审美观的影响，形成了传统的几何形、图案式风格。宫廷插花常以大口径圆罐作容器，以草本花卉为主要材料，造型简单规整，花朵均称丰满，色彩艳丽丰富，体量较大。民间插花则以欢快、简朴的自由插花风格为主。

19世纪下半叶是西方家庭园艺和西方传统插花的黄金时期，插花成为时尚，用插花装饰餐桌及居室已成为文明风雅的生活艺术。西方插花逐渐走向理论化、系统化，插花作品色彩浓烈，花材量大，以几何构图为主，严格要求对称和平衡，层次分明有规律，表现出一定的节奏，以人工美取胜，充分表现了欧洲传统插花的特点。

With the influx of mass fresh flowers into Japan after the war, new concepts of floral material and shaping evolved. Fashion-forward flower arrangements turned up, that was the modern flower arrangement focusing on shaping.

1.2.3 A Brief history of Occidental Flower Arrangement

Flower arrangement in Western countries originated in ancient Egypt and Greece. Egyptians had *Nymphaea tetragona* inserted in bottles or bowls as decorations, gifts or funerals for customs. Along with the spread of culture, flower arrangement had successively been spread to Belgium, Holland, Britain and the like.

In early stage, flower arrangement was spread in Europe as religious flower arrangement. Westerners believed that flowers could expel witchcraft and lightning, and often used olive branches and laurel branches as wreaths worn around the head or neck for protective talismans or being hung on doors, walls to prevent evil spirit from entering rooms.

During the 14th~16th century — the European Renaissance Period, flower arrangement got rid of religion and rapidly developed. The style was influenced by geometrical aesthetics in western art, forming a traditional geometric pattern. Round cans with large caliber were usually used as vessels in palace flower arrangement. Herbal flowers were chosen as main materials, structures were simple, clear and neat, flowers were symmetrical and full, colors were gorgeous and rich, and the volume was large. On the other hand, the style of flower arrangement among the folk featured hilarious, free and simple.

The second half of 19th century was the golden period of Western family gardening and their traditional flower arrangement. Flower arrangement had come into fashion. Decorating table and living rooms with flower arrangement had become civilized and elegant living art. Western flower arrangement gradually became theoretical and systematic. The features of flower arranging works were of colors, large quantity, geometric composition and strict requirement

第二次世界大战后，日本花道传入欧美，东西方插花相互融合，形成了更具时代感和艺术魅力的现代西方插花。近代由于设计学、构成学的介入，西方现代花艺更是日新月异、蓬勃发展。

1.3 插花艺术的分类

插花艺术种类繁多，可从不同角度加以分类，这里介绍几种常见的分类方法。

1.3.1 按民族风格分类

世界上插花流派纷呈，风格各异。由于文化传统、风俗习惯、哲学信仰等不同，形成了风格截然不同的东西方两大插花流派。

（1）东方插花

东方插花以中国、日本为代表。作品以儒雅、清新、简练、自然为特征，重在姿韵和意境的表现，以神取胜。

（2）西方插花

西方插花以欧美为代表。作品以华丽、热烈、丰满、规则为特征，重在色彩和装饰的表现，以盛取胜。

1.3.2 按时代分类

插花艺术和其他艺术一样，随着时代变迁而发展。插花艺术分为古代插花和现代插花，它们又各有不同的分类。

for symmetry and balance, showing a certain rhythm, winning for artificial beauty and fully demonstrating the characteristics of European traditional flower arrangement.

After World War II, Japan Ikebana was introduced to Europe and America, Oriental and Western flower arrangement were merged into each other and formed a more modern and artistic charming western flower arrangement. In modern times, due to the intervention of design science and construction science, western modern floriculture is changing and flourishing more rapidly.

1.3 Classification of Flower Arrangement Art

There are many different kinds of flower arrangement art, which can be classified from different angles. Here are some common classification methods.

1.3.1 Classification According to National Styles

There are many different flower arrangement schools with different styles. Because of different cultural traditions, customs, habits, and philosophical beliefs, Eastern and Western flower arrangement schools with distinctly different styles have been formed.

(1) Oriental Flower Arrangement

China and Japan are two representatives of Oriental flower arrangement. Characteristics of those works are elegant, clear, concise and natural, focusing on the expression of charm and artistic conception, winning by spirit.

(2) Western Flower Arrangement

Europe and America are the representatives of Western flower arrangement. Characteristics of those works are gorgeous, warm, plump and regular, focusing on the expression of color and decoration, winning by magnificence.

1.3.2 Classification According to Ages

Like other arts, flower arrangement art developed as time changed. Flower arrangement arts can be classified into ancient flower arrangement and modern flower arrangement.

（1）古代插花

①宗教插花：宗教插花主要用于佛堂供佛，也称供花。供花插花形式不讲究，花材多选择睡莲、荷花等清净植物，显得庄严肃穆。

②宫廷插花：宫廷插花结构严谨、讲究排场，主要用于宫廷装饰，也称院体花。宫廷插花形体硕大、色彩华丽、枝叶繁茂、装饰趣味浓厚。有的用十种花材插制而成，寓意十全十美。有的取其谐音，求其幸福吉祥之意，如将玉兰、海棠、牡丹插在一起，寓意玉堂富贵。宫廷插花选用牡丹、芍药、山茶、梅、竹、松等花材较多，显得富丽堂皇，寓意吉祥。

③民间插花：民间插花大多在节庆期间兴起，如逢春节、元宵、端午、七夕、中秋等节日，人们摆上一盆盆时令插花，并以春联、剪纸、花结、年画、花灯、香包等配合装饰，烘托出喜气洋洋、欢乐满堂的节庆气氛。民间插花不讲究造型，随兴所至，但讲求吉祥、热闹，多数采用时令色彩鲜艳的花材插制，显得热闹喜气。因其插制随意，又称自由插花。

Each of them has different classifications.

(1) Ancient Flower Arrangement

① Religious flower arrangement: Religious flower arrangement is mainly used for worshiping Buddha in temples, which is also known as flower offering. The style is free, and clear plants such as *Nymphaea tetragona* and *Nelumbo nucifera* are often selected as materials, looking solemn.

② Palace flower arrangement: Palace flower arrangement has rigorous structure, paying attention to ostentation and extravagance, which is mainly used for palace decorations, also known as flower arrangement in courtyard. Its shape is huge with gorgeous color, luxuriant foliage and strong decorative taste. Some are arranged with ten flower materials, meaning perfection. And some take their homophonic implications, meaning happiness and luck, for example, *Magnolia denudate* (magnolia), *Malus spectabilis* and *Peony suffruticosa* are arranged together implying riches and honor. The materials of flower arrangement in palace are *Peony suffruticosa* (Camellia), *Camellia japonica* (camellia), *Armeniaca mume*, Bambusoideae (bamboo), *Pinus* spp. and so on, appearing splendid and auspicious.

③ Folk flower arrangement: Most folk flower arrangements rose during festivals, such as the Spring Festival, the Lantern Festival, the Dragon Boat Festival, the Chinese Valentine's Day, the Mid-Autumn Festival and so on. People often put seasonal flower arrangement in pots with the Spring Festival couplets, paper-cuts, knots, the Chinese New Year pictures, lanterns, sachets, and other decorations to add beaming/jubilant and joyful festival atmosphere. The styles of folk flower arrangement are free, according to personal interest, but being particular about luck and jollification. Most are arranged with brightly colorful flowers, appearing lively and festive. Because of free arrangement, it is also known as free flower arrangement.

④文人插花：与宗教、宫廷、民间插花三种风格迥然不同的是文人插花。文人插花脱俗清雅，它不重排场、不为祈福，主要讲究情趣，借花明志抒情，又称心象花。所用花器较为朴实，花材以素雅的兰、竹、梅、菊、百合等为主。由于一批画家、书法家、诗人、文学家的介入，文人插花开始注意形式美的法则，讲究比例结构和平衡韵律，这为插花作品注入了丰富内涵，达到了形式美和意境美的圆满结合。文人插花使中国插花艺术步入了专门艺术。

（2）现代插花

①礼仪插花：礼仪插花顾名思义是指在礼仪场合和人际交往中使用的插花，如花篮、花束、花钵、花车、花圈等，是用于节庆庆典、婚宴、迎宾、乔迁、庆贺生日、舞台献花、探望病人、悼念故人以及各种公共场所的花饰。礼仪插花通常有比较固定的形式，用花量较多，侧重装饰效果和感情效果。由于礼仪插花商品性强，又称商业插花。

②大众插花：大众插花指民众把花买回家后，随意插制，加以欣赏。它并不特别讲究构图造型或意境内涵，主要是通过花的色彩和芳香，装饰环境、增添生气。

④ Flower arrangement by scholars: Flower arrangement by scholars is completely different from all the above. Its features are refined and elegant, without paying attention to ostentation and extravagance, not for blessing, mainly striving for individual interest, taking it voicing their ambitions. Flower arrangement by scholars is a style namcd "mood flower". Vases are simple, floral materials are mainly elegant *Cymbidium* ssp., Bambusoideae, *Pinus* spp., *Dendranthema morifolium*, and *Lilium* spp. (lilies). Because of the involvement of some painters, calligraphers, poets and litterateurs, scholars began to notice the rules of beauty in form, structure, and balanced rhythms, which pour rich connotation into those works to successfully combine formal beauty and artistic conception beauty together. Flower arrangement by scholar makes Chinese flower arrangement evolve into a professional art.

(2) Modern Flower Arrangement

① Etiquette flower arrangement: Etiquette flower arrangement refers to flower arrangement used on ceremonial occasions and interpersonal relationships. For example, flower basket, bouquet, flowerpot, float, and wreath and so on are mainly used for festival celebration, wedding reception, greeting guests, moving to a better place, celebrating birthday, offering bouquets on the stage, visiting the sick, mourning friends, and in various public places. Etiquette Flower Arrangement has a fixed form, using plenty of flowers, focusing on decoration effects and emotional effects. Because etiquette flower arrangement has the characteristic of commodity, it is also called commercial flower arrangement.

② Public flower arrangement: Public flower arrangement refers to random flower arranging to appreciate them after people buy flowers and take them home. It is not particular about exquisite composition, modeling, and connotation, decorating surroundings and enhancing lively atmosphere by the color and aromas of flowers.

③艺术插花：艺术插花既重形式又重意境，它由古代文人插花发展而来。艺术插花有写意插花、写景插花和抽象插花。艺术插花形式多变，不拘一格，用花量比礼仪插花少，但造型讲究，内涵丰富，又称创意插花。

④现代花艺设计：现代花艺设计也可理解为广义的艺术插花，作品有的有器皿，有的无器皿；有的置于台桌，有的置于地面或空间；有的装饰环境，有的装饰人体。花艺设计的特点包括：

空间性：花艺作品可以布置在任何空间位置，有落地的，有悬挂的；有固定的，有流动的；有室内的，有室外的等。

组合性：现代花艺技术通常体量较大，它往往由几个单体插花作品组成，有几个焦点，但又组合成一个整体。

立体性：花艺作品比插花作品更具立体性，它需要在高低、前后、左右全方位展开，三维空间处理更为细微。有些大型作品甚至可进入观赏。

独创性：现代花艺作品往往不运用常规器皿，花器与花材连成一体，由设计者自行设计创作，个性化突出，很少雷同，增强了作品的独创性。

综合性：现代花艺设计是插花艺术发展到一定高度的产物，它是插花艺术与其他门类艺术的结合，如建筑艺术、装潢艺术、园林艺术、舞台艺术、橱窗艺术、工艺美术等，也是艺术和技术的结合，如构图学、色彩学、文学、植物学、材料学、力学、工程学、光学等。

③ Artistic flower arrangement: Artistic flower arrangement focuses on both form and artistic conception. It inherited ancient scholars' flower arrangement and then developed. It includes freehand flower arrangement, scenery flower arrangement and abstract flower arrangement. The forms of artistic flower arrangement are various, without sticking to one pattern. The number of flowers needed is less than that of etiquette flower arrangement. But it pays great attention to modeling and has rich connotation, so it is also called creative flower arrangement.

④ Modern floral design: Modern floral design can also be understood as broad artistic flower arrangement. Some works have vessels, others don't have; some are put on the desk, others are put on the ground or space; some are used to decorate surroundings, others are used to decorate human body. Features of floral design are as follows:

Spatiality: Floral works can be arranged in any spatial location, with landing, hanging; being fixed, flowing; being indoor or outdoor, and so on.

Combination: The body of flower arrangement with modern technique is often large, which is made of several single flower arrangement works with several focuses but being combined into a whole.

Three-dimension character: Floral works are more three dimensional than flower arrangement's. The works expand in all directions, high or low, front or back, left or right, three-dimension space is more subtle. Some large-scale works even can allow people to enter for appreciation.

Originality: Conventional vessels are not often used in modern floral works. Flower vessels are connected with flower materials as a whole. Designers design works on their own, stressing individual character, and the works are rarely identical, strengthening its originality.

Comprehensiveness: Modern floral works design is emerged with the development of flower arrangement art to a higher stage. It is a combination of flower arrangement art and other kinds of arts, such as architectural art, decoration art, horticulture, stage art, show window art, crafts, etc.

所以，大中型现代花艺创作要求很高，需要在一般插花基础上，增加更多的知识。

1.3.3 按器具分类

（1）瓶花

瓶花是古今中外最早出现、最早使用的插花形式。花瓶是竖立式容器，形态有直筒形、小口大肚形、大口小底形、葫芦形、球形、长方形、方形等。质地有陶、瓷、铜、景泰蓝、竹编、藤编、玻璃、塑料、玻璃钢等。

（2）盆花

盆花也是古老而普遍应用的插花形式。花盆是宽口、浅身、扁平的容器。形态有圆形、椭圆形、半圆形、方形、长方形、梭形、荷叶边形等。质地有陶瓷、瓷、漆器、石、景泰蓝、竹编、藤编、玻璃、塑料、玻璃钢等。

（3）篮花

花篮插花常见于礼仪插花中，但也经常见诸艺术插花中，作为纯观赏的艺术品。花篮通常用竹、藤、柳、麦秆等编制，也有用陶瓷、玻璃等材料制作。形态有有柄花篮、无柄花篮、筒状花篮、提篮式花篮、酒坛状花篮、船形花篮、壁挂花篮、平底状花篮、月亮形花篮等。

It is also a combination of art and technique, such as composition, color science, literature, botany, material science, mechanics, engineering science, optics, etc.

Therefore, large or medium-sized modern floral art design has higher requirements for creation. And designers need to master more knowledge on the basis of general flower arrangement knowledge.

1.3.3 Classification According to Equipment

(1) In vase

It is the earliest flower arranging form and used at all times and in all places around the world. Its vase is an erected container, its shapes are straight type, narrow-necked and pot-bellied type, wide-necked and small bottom type, pear-shaped, spherical, rectangular, square, and so on. The materials of the vases are made of ceramic, porcelain, bronze, cloisonne, bamboo weaving, rattan weaving, glass, plastic, and fiberglass, etc.

(2) In pot

It is also an ancient and widely-used form of flower arrangement. The pot is a wide, shallow, and flat container. The shapes are spherical, oval, square, rectangle, fusiform, flouncing and so on. The materials are made of ceramic, porcelain, lacquerware, stone, cloisonne, bamboo weaving, rattan weaving, glass, plastic, fiberglass, etc.

(3) In basket

It is usually used in etiquette flower arrangement, sometimes also used in other flower arrangement arts as a purely ornamental art. The baskets are often woven of bamboo, rattan, willow, straw, and made of ceramics, glass and other materials. The shapes are petiolate flower baskets, sessile flower baskets, tubular-shaped flower baskets, basket-shaped flower baskets, jar-shaped flower baskets, boat-shaped flower baskets, wall-hanging flower baskets, flat bottom flower baskets, moon-shaped flower baskets, etc.

(4) In tubular

Bamboo tubular flowers are representatives of tubular

（4）筒花

筒花以各式竹筒花为代表，后又有仿照竹筒的陶瓷筒、玻璃筒、藤编筒、木筒等。

（5）钵花

钵花的器皿为低矮粗圆的花器，它比花瓶低而粗，比花盆高而瘦，显得粗犷稳重。宜插制落地式插花或花材数量多的插花。其外形多为缸状，能储水。材质以陶瓷为主，也有用石膏、玻璃钢等材料制成的。

（6）碗花

碗花的器皿为口宽底窄的圆形花器，在中国古代和日本花道中运用较多，现代插花中往往把它归为盆花。

除了上述6种花器以外，现代尚有似盆非盆、似瓶非瓶，形状抽象独特的各种花器，称为异形花器。异形花器的出现为艺术插花的创作拓宽了变化空间。

1.3.4 按表现手法分类

（1）写景插花

写景插花是把大自然中某一局部，通过艺术加工再现于瓶盆之中。创作要求源于自然、高于自然。通常使用浅口大盆插制，或表现庭园一隅，或表现池塘水景，或表现山野一角。

（2）写意插花

作品注重写实景致，重在意蕴。作者往往借助于器皿、花材、造型、色彩等方面元素，抒发各人心志、颂扬社会美德、诉说宇宙哲理等。

（3）抽象插花

抽象插花造型不拘一格，花材可以重新分解组合，并通过各种新的表现手法，达到不同凡响的艺术效果。

flowers. Next to that, there are ceramic tubes, glass tubes, rattan-weaving tubes, wooden tubes, and so on, which imitate tubular flowers.

(5) In farthen bowl

Its vessels are low, thick and round. It is lower and thicker than flower vases, and higher and thinner than pots, looking rough and steady. It's suitable for console flower arrangement or inserting a lot of flowers. The shapes look like vats which can store water. The main material is ceramic, gypsum, fiberglass and other materials are also used.

(6) In bowl

Its vessels are wide-necked, narrow bottom round flower vessels, usually used in Chinese ancient times and Japan ikebana. It also belongs to pot flowers in modern flower arrangement.

In addition to the above six kinds of flower vessels, there are various kinds of abstract unique flower vessels which look like pots and vases and are called alien vases. The appearance of alien vases widens variation space for the creation of artistic flower arrangement.

1.3.4 Classification according to Expression Technique

(1) Flower arrangement in scenery

It means that we reproduce a part of nature in flower organs through artistic processing. Creations require that the resources stem from nature and be higher than nature. It is usually arranged in shallow basins, performs a corner of garden, or pond scenery, or a corner of the mountain.

(2) Freehand flower arrangement

Its works focus on realistic scenery and implication. Authors often use vessels, materials, and modeling, colors and other to express their ambitions and emotions, praising social virtues and telling universal philosophy and so on.

(3) Abstract flower arrangement

Its modeling is not limited to one type, flower materials can be disintegrated and combined. Through various new performances, it can achieve remarkable artistic effects.

抽象插花有注重装饰效果的，也有装饰和意境两者兼备的。

1.3.5 按使用场合分类

（1）家庭插花

家庭插花有玄关插花、客厅插花、餐厅插花、卧室插花、书房插花、过道插花、阳台插花等。

（2）宾馆插花

宾馆插花有大堂插花、总台插花、客厅插花、卧室插花、琴房插花、餐厅插花、酒吧插花、阳台插花等。

（3）婚庆插花

婚庆插花有新娘花饰、花车、花束、宴席环境插花、餐桌插花、椅背插花、新房插花等。

（4）会场插花

会场插花有入口插花、会议桌插花、休息厅和接待厅插花、环境插花等。

（5）写字楼插花

写字楼插花有大堂插花、会客厅插花、会议室插花、办公室插花、过道插花等。

（6）商厦插花

商厦插花有大门花艺、橱窗花艺、商铺插花、接待室插花、休息室插花、办公室插花、餐厅插花、咖啡吧插花等。

（7）丧用插花

丧用插花有灵堂插花、灵柩插花、悼念花篮、休息室插花等。

Abstract flower arrangement focuses on decorative effects, or both decoration and artistic conception.

1.3.5 Classification According to Application

(1) Family flower arrangement

There are entryway flower arrangement, living room flower arrangement, dining room flower arrangement, bedroom flower arrangement, study room flower arrangement, aisle flower arrangement, balcony flower arrangement and so on.

(2) Hotel flower arrangement

There are lobby flower arrangement, reception desk flower arrangement, living flower arrangement, bedroom flower arrangement, Piano room flower arrangement, dining room flower arrangement, bar flower arrangement, balcony flower arrangement and so on.

(3) Wedding flower arrangement

There are bride floral ornamental designs, floats, bouquets, flower arrangement for dinner environment, table flower arrangement, chair flower arrangement, flower arrangement in bridal chamber and so on.

(4) Flower arrangement in meeting place

There are entrance flower arrangement, conference table flower arrangement, lounge flower arrangement, reception hall flower arrangement, environmental flower arrangement and so on.

(5) Flower arrangement in office building

There are hall/lobby flower arrangement, reception room flower arrangement, conference room flower arrangement, office flower arrangement, aisle flower arrangement and so on.

(6) Flower arrangement in commercial building/mall

There are doorway floral arrangement, show window floral arrangement, shop flower arrangement, reception room flower arrangement, waiting room flower arrangement, dining room flower arrangement, coffee bar flower arrangement and so on.

(7) Flower arrangement for funeral affairs

There are mourning hall flower arrangement, bier

1.4 插花艺术的作用

1.4.1 环境作用

（1）柔化空间

人们在生活的空间中，接触建筑、家具等硬线条实物较多，视觉较为单一、生硬。布置线条优美的插花作品，能使人们的视觉得到缓冲，感到赏心悦目。

（2）生化环境

人们在工作或生活环境里接触无机物较多。布置有生命的插花作品，能增添生气。尤其到了冬天，室外万物萧条，室内春意盎然，带给人们生机勃勃的感受。

（3）优化生活

插花是高雅艺术，是家庭或工作场所的软装潢。在工作或生活的环境里，布置几件插花作品，能保护视力，调节心情，享受自然。

1.4.2 文化作用

（1）传承祖国优秀文化

中国插花艺术是东方插花的基础和重要组成部分，也是我国的"国粹"。普及和提高中国插花艺术有益于弘扬祖国优秀传统文化，发展健康有益的民族文化。

（2）陶冶情操

插花是一种追求真、善、美的高雅文化活动，无论是创作插花还是欣赏插花，插花形态美和意境美的熏陶，有利于培养人们高雅的生活情趣，追求高尚的行为准则。

flower arrangement, mourning flower arrangement, lounge flower arrangement and so on.

1.4 Functions of Art of Flower Arrangement

1.4.1 Environmental Function

(1) Softening space

People are in touch with more hard-edged objects such as architecture, furniture and others in their living space. The vision is single and blunt. Decorating graceful flower arrangement can make people's vision pleasant and buffering.

(2) Enlivening environment

In working or living environment, people are exposed to many inorganic substances. Arranging some living flower arrangement works can add vitality to environments. Especially in winter, outdoor environments are in all depression, Indoor environments are furious spring, which bring people a vibrant feeling.

(3) Optimizing life

Flower arrangement is a refined art and a soft decoration of family and working places. In working or living environments, arranging several flower arrangement works can protect eyes, regulate mood and enjoy nature.

1.4.2 Cultural Function

(1) Inheriting national excellent culture

Chinese flower arrangement art is the basic and important part of the Oriental flower arrangement. It is also the "Quintessence" of Chinese culture. Popularity and improving Chinese flower arrangement art is good for carrying forward our national excellent traditional culture and developing healthy and beneficial national culture.

(2) Cultivating mind

Flower arrangement is an elegant cultural activity that pursues the truth, the goodness, and the beauty. Both creation of flower arrangement and appreciation of flower

（3）提高文化艺术素养

人们在学习和创作插花艺术过程中，会不断丰富构图色彩、花卉植物、文学诗歌等知识。日积月累，插花水平提高了，文化艺术素养也会相应提高。

1.4.3 感情作用

花卉是世界上最美好的象征物，人们在不同的节日或特定的时间场合，通过送花来传递友情、亲情、爱情乃至爱国之情。花卉在人们感情生活中起了无法替代的润滑剂作用，可以说插花是服务于人们昨天、今天、明天的感情使者。

1.4.4 经济作用

插花引导鲜切花消费，鲜切花消费促进鲜切花的经营和生产。插花有利于调整农业结构和增加农民收入。同时，由于插花逐渐成为一个产业，它也有利于城市再就业。

插花除了使用鲜花外，还需要盆、瓶、篮、筒等器皿，各种工具和装饰物。插花活动的发展，也必然带来相关资材的发展，促进相关产业。

另外，仿真花历来是中国的优秀工艺品。中国的人造花，其仿真程度之高超，制作技艺之精到，令各国插花花艺家赞叹不已。随着插花活动的普及，仿真花销量也与日俱增，它必然带来经济效益和社会效益。

arrangement are beneficial for cultivating people's graceful living interests and pursuing noble behavior standards through flower arranging formal beauty and edification of artistic conception beauty.

(3) Improving cultural and artistic accomplishment

In the process of learning and creating flower arrangement art, people will continuously enrich composition, colors, flowers, plants, knowledge of literature and poems and so on. With the accumulation over time, flower arranging levels are improved, and cultural artistic accomplishment will be improved accordingly.

1.4.3 Emotional Function

Flower is the most beautiful symbol in the world. People send flowers to express their friendship, affection, love and even universal love for the whole country in different festivals or at a special time. Flower plays an irreplaceable part in people's emotional life as lubricant. It is flowers that serve people as emotional messengers yesterday, today and tomorrow.

1.4.4 Economic Function

Flower arrangement guides the consumption of fresh cut flowers, which promotes its operation and production. Flower arrangement is beneficial to adjust agricultural structure and increase peasants' income. At the same time, because flower arrangement is becoming an industry, it is also in favor of the urban reemployment.

Besides the use of fresh flowers, flower vessels such as basins, bottles, baskets, tubes, and so on, various tools and ornaments are also needed. The development of flower arrangement activities will surely bring related materials' development to promote related industries.

Artificial flowers have always been Chinese excellent art crafts. Flower arranging artists from different countries highly praised Chinese artificial flowers with high level simulation and exquisite production techniques. With the popularity of flower arranging activities, artificial flowers'

插花是一种创意产业，它是环境艺术设计的组成部分。发展插花事业有利于发展创意产业。

1.5 插花艺术评判标准

欣赏和评判插花作品的优劣成败，通常从以下 4 个要素入手。

1.5.1 构图造型

造型、器具、花材要符合主题，统一协调。

比例合适，三维空间明显。

作品要达到视觉上的平衡。

作品要具有创新和美感。

1.5.2 色彩配置

花器、花材、配件的色彩要符合主题。

各色彩之间过渡自然、协调美观。

处理好色彩的整体感和平衡感。

1.5.3 主题意境

切题：情真意切、令人信服。

健康：颂扬真、善、美，激励人生。

含蓄：有文化品位，启人遐思，耐人寻味。

sales also grow increasingly, it will inevitably bring economic and social benefits.

Flower arrangement is a creative industry and a component of environmental artistic design. The development of flower arrangement is beneficial to the development of creative industries.

1.5 Criteria for Appraising Flower Arrangement Art

The ways of telling the advantages and disadvantages of flower arrangement works when appreciating and evaluating generally start with the following four elements.

1.5.1 Composition and Plastic Art

Modeling, vessels and materials should be in accordance with the theme and achieve unity and harmony.

Proportion should be right, and the three-dimension space is obvious.

Works should achieve visual balance.

Works should have innovation and aesthetic feeling.

1.5.2 Color Scheme

The colors of flower vessels, materials and accessories should be consistent with the theme.

The transition of colors should be natural, coordinating and beautiful.

The overall sense of color and sense of balance should be handled properly.

1.5.3 Theme of Artistic Conception

Keeping to the point: being sincere and convincing.

Being healthy: praising the truth, the goodness, and the beauty and inspiring life.

Being implicit: having cultural taste, making people think, and being durable for reflection.

1.5.4 制作技巧

插制前,有序置放花材、花器和制作工具。

花材选择要优质新鲜,对花材的处理要修剪合理、加工精细。

处理好花材的固定和作品的稳固性。

遮盖处理自然,不露人工痕迹。

花束、胸花、肩花、头花等制作,以及绑扎物要求对人体安全。要遮盖整洁,外形美观。

作品完成后,清理桌面和现场。

这4个评判标准,也是插花员需要牢牢把握好的创作要素。

1.5.4 Producing Skills

Before arranging, put flower materials, flower vessels and tools in order.

Select high-quality fresh flowers, trim flower materials properly and process finely.

Fix flower materials steadily and make them firm.

The work of covering should be natural without any artificial traces.

The production of bouquets, flower brooches, flowers on shoulders, headdress flowers and lashings should be safe for people. Cover neatly and make appearance beautiful.

After the works being finished, clean the desk and the spot.

Those four criteria above are also the creating elements for flower arranging staff to grasp firmly.

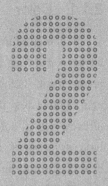

第2章 花材花器

Chapter 2　Flower Materials and Conventional Equipment for Flower Arrangement

2.1 花卉的色彩和形状

2.1.1 花卉的色彩

色彩是构成美的重要因素,西方插花特别强调色彩的应用。俗话说:"远看色彩近看花",花材本身色彩鲜艳丰富,使用基本的色彩知识进行花色搭配,能够创造出具有形式美、内在美、意境美的优秀插花作品。

(1)红色系花材

红色系花材是指花卉的主要观赏部位颜色是红色系。红色系花材的花材有紫薇、红杏、山茶、香石竹、月季、玫瑰、一品红、贴梗海棠、梅等。红色是生命、活力、健康、热情、朝气、欢乐的象征。由于红色在可见光谱中光波最长,所以最为醒目,给人视觉上一种迫近感和扩张感,容易引发兴奋、激动、紧张的情绪。运用红色系花材插制的插花作品可以营造一种喜庆、热烈、欢快的氛围。

(2)黄色系花材

黄色系花材是指花卉的主要观赏部位颜色是黄色系。黄色花系的鲜花有向日葵、大丽菊、黄菊、香石竹、萱草、美人蕉、黄水仙、蜡梅、郁金香、旱金莲、月季等。黄色是富贵的象征,古代宫殿建筑和器物都是黄色的,清代皇帝以黄马褂赐给立功的高官。但黄色同时在丧礼上普遍应用。在日本,黄菊只应用于丧礼;西方送黄玫瑰表示分手。与黄色较相似的橙色,体现了光辉、温暖、欢乐、热烈的情绪,同时是尊严、高贵、权势、富裕的象征。

2.1 Color and Shape of Flower

2.1.1 Color

Color is an important element in beauty constitution. The application of color is especially emphasized in the Western flower arrangement. As the saying goes: "Appreciating the colors from afar, appreciating the flowers nearby." Floral material itself is already rich in color. So by using basic color knowledge to have assortment, an outstanding art arrangement with the beauty of form, inner and artistic conception can be created.

(1) The symbol of red flower

Red flowers refer to those with the mainly red ornamental parts. Such floral materials are from *Lagerstroemia indica*(crape myrtle), *Armeniaca vulgaris*(red apricot), *Camellia japonica*(camellia japonica), *Dianthus caryophyllus* (carnation), *Rosa chinensis* (Chinese rose), *Rosa rugosa* (rose), *Euphorbia pulcherrima* (poinsettia), *Chaenomeles speciosa*(chaenomeles speciosa) and *Armeniaca mume*. Red color is the symbol of vitality, energy, health, passion, youth and happiness. As the red light in the visible spectrum is the longest, it is so arresting that it can give a vision sense of pressure and expansion which are inclined to stimulate the emotion of excitement and tension. Artistic arrangement with red floral materials can create a festive, lively and cheerful atmosphere.

(2) The symbol of yellow flower

Yellow flowers refer to those with the mainly yellow ornamental parts. Such floral materials are from *Helianthus annuus* (sunflowers), *Dahlia pinnata* (Dalia pinnata), *Dendranthema indicum* (chrysanthemum), *Dianthus caryophyllus* (carnation), *Hemerocallis fulva* (daylily), *Canna indica* (India canna), *Narcissus pseudonarcissus* (daffodils), *Chimonanthus praecox*, *Tulipa gesneriana*(tulips), *Tropaeolum majus* (tropaeolum majus) and *Rosa chinensis* . Yellow is the symbol of wealth, so ancient palace buildings and artifacts are yellow; the emperor of the Qing Dynasty even gave the yellow jacket to meritorious officials as an award. However, the color of yellow is universally applied in

插花作品中运用黄色或橙色花材可以营造一种温暖、热情、大气的氛围。

（3）蓝紫色系花材

蓝紫色系花材指花卉的主要观赏部位颜色是蓝紫色系。蓝紫色花系的鲜花种类包括紫罗兰、蝴蝶花、菊花、紫玉兰、紫藤等。蓝色会让人联想起大海、星空，使人安静，有深远和清新的感觉，但也使人产生阴郁、贫寒和冷淡之感。比利时人最忌蓝色，不吉利的场合都穿蓝色衣服。但有时蓝色会成为时尚流行色，视当年流行情况而定。插花作品中运用蓝紫色花材可以营造一种宁静、沉稳的氛围。

（4）白色系花材

白色系花材是指花卉的主要观赏部位颜色是白色系。白色花系的鲜花包括梅花、百合、玉兰、李、栀子、茉莉、菊花、晚香玉、白月季、白色马蹄莲（增加种类）。白色的鲜花象征纯洁、无暇、高尚、神圣、清净、素雅、善良。常见的西方式婚礼用花多以白色为主体，体现婚礼的神圣和新娘的纯洁。

（5）其他颜色花材

绿色系花材指绿色的叶材、绿色的鲜花。绿色的叶材最普遍，但绿色的鲜花品种很少，如菊花品种碧绿油绿，绿色香石竹，绿色洋桔梗。绿色富有生机，富有春天气息，使人安静、沉着和理智。嫩绿色尤其使人感受到新生命的活力和希望。

the funerals. In Japan, *Dendranthema indicum* is used only in funerals; in the west, *Rosa hybrida*(yellow rose) represents breaking-up of lovers. The color of orange which resembles yellow reflects a brilliant, warm, happy and enthusiastic mood, and it is also a symbol of dignity, nobility, power and wealth. Artistic arrangement with yellow and orange floral materials can create a warm, passionate and meteoric atmosphere.

(3) The symbol of bluish violet flower

Bluish violet flowers refer to those with the mainly bluish violet ornamental parts, which contains *Matthiola indica*, *Iris japonica* (iris), *Dendranthema mordifolium*, *Magnolia liliflora* and *Wisteria sinensis* (Chinese wisteria). The color of blue always reminds us of the sea and the sky, makes us quiet, gives us the sense of being profound and refreshed, whereas it also fills us with the sense of gloom, poverty and coldness. For instance, Belgians are so sensitive to blue color that they always wear blue clothes on some unlucky occasion. However, blue color can become popular sometimes, which is determined by the popular trend of the very year. Artistic arrangement with bluish violet floral materials can create a quiet and calm atmosphere.

(4) The symbol of white flower

White flowers refer to those with the mainly white ornamental parts, which include *Armeniaca mume*, *Lily* spp., *Magnolia denudata*, *Prunus salicina*, *Gardenia jasminoides*, *Jasminum sambac* (jasmine), *Dendranthema morifolium*, *Polianthes tuberosa* (tuberose), *Rosa chinensis* (white Chinese rose), and *Zantedeschia aethiopica* (white calla). White flowers are the symbol of cleanliness, pureness, nobility, holiness, simplification, elegance and kindness. Common Western weddings prefer white flowers as decoration to represent the bride's purity and the holiness of wedding.

(5) Flowers of other colors

Green flowers refer to those with the green leaf materials and flowers. Green leaves are very common, but green flowers are so rare in nature that we can only find *Dendranthema morifolium* 'green' and *D. morifolium* 'clumps', green *Dianthus caryophyllus*, green *Platycodon grandiflorus* (balloon flowers). Green color is full of vitality and the scent

黑色系花材很少见。实际上是没有纯黑花卉的,只是浓重的紫色产生黑色的感觉。主要的黑色系花材有黑牡丹、黑郁金香和墨菊。黑色鲜花使人产生神秘、庄重的感觉。

2.1.2 花卉的形状

(1) 线状花材

线状花材指花的外形为长条状、花序呈线状排列的花材。线状花材可构成插花作品主体结构,是构成花型轮廓和基本构架的主要花材。如具有总状花序、穗状花序、柔荑花序、肉穗花序的各种花卉。此类花材有唐菖蒲(剑兰)、金鱼草、紫罗兰、晚香玉、紫罗兰、蛇鞭菊、飞燕草、贝壳花等。此外,各种木本植物的枝条、根、茎、长形叶、芽和藤蔓植物也可作为线状花材。

(2) 团块状花材

团块状花材指花朵呈圆球形或圆盘形、花形对称的花材,常作为焦点花。插花作品的主体常用团状花材完成。团状花材种类最多,主要有月季、芍药、香石竹、菊花、非洲菊、向日葵、郁金香、茶花、百合、睡莲等。

(3) 特殊形状花材

特殊形状花材一般指花形奇特的花材,在插花作品中常作为焦点花应用,一般放在插花作品重心的位置或作品最醒目的位置。如马蹄莲、花烛、天堂鸟、蝴蝶兰、卡特兰、鸡冠花等。特殊形状的花材一般较贵重。

of spring, which will make people quiet, calm, cool and reasonable. Peak green, in particular, can make people gain the vitality and hope of new life.

Black flowers are so rare that actually, there is no absolutely black flower. It is just dark purple that gives an illusion of black flowers. Black flowers mainly include black *Peony suffruticosa*, black Tulipa gesneriana and black Dendranthema morifolium. Black flowers can create a feeling of mystery and solemnity.

2.1.2 Flower's Shapes

(1) Lineal flower material

Linear flower materials refer to those with flowers shaped long and stripped, and with the inflorescence arranged lineally, which constitute the main structure of the flower arrangement and are also the main flower materials to constitute the outline and basic framework of the flower arrangement. Just like various flowers with racemes, spice, catkin and spandex, these types of floral materials include *Gladiolus gandavensis* (gladiola), *Antirrhinum majus* (snapdragon), *Matthiola incana* (violets), *Polianthes tuberosa*, *Liatris spicata* (liatris spicata), *Consolida ajacis* (delphinium ajacis), and *Moluccella laevis* (molucca balm). In addition, a variety of ligneous branches, roots, stems, long leaves, buds and vine plants can also be used as linear floral materials.

(2) Crumby/cloddy flower material

Crumby flower materials refer to those with spherical or disc flowers and symmetrical flower types, which are often used as the focus flowers to constitute the main element of flower arrangement. Of the different flower materials, crumby flower materials has the most sources, such as *Rosa chinensis*, *Paeonia suffruticosa*, *Dianthus caryophyllus*, *Dendranthema morifolium*, *Gerbera jamesonii*, *Helianthus annuus*, *Tulipa gesneriana*, *Camellia japonica*, *Lilium* spp., and *Nymphaea tetragona*.

(3) Specially shaped flower material

Specially shaped flower materials usually refer to flowers with fancy shapes, which are often used as the

（4）点缀填充花材

点缀填充花材指花形细碎小巧、散开分布或密聚成束、呈伞状的花材，主要作为填充花，修饰作品整体造型，配合主花使用。常见点缀填充花材有满天星、情人草、勿忘我、小菊、孔雀草、一枝黄花、文竹、天门冬等。

2.2 常用草本花材及花语

（1）草本观花花材（表2-1）
（2）草本观叶花材（表2-2）

2.3 常用木本花材

（1）木本观枝、观花花材（表2-3）
（2）木本观叶花材（表2-4）
（3）木本观果花材（表2-5）

focus flowers placed in barycentre or the most striking position of flower arrangement. Such materials are *Zantedeschia aethiopica*, *Anthurium* spp. (anthurium andraeanum lind), *Strelitzia reginae* (bird of paradise), *Phalaenopsis aphrodite* (phalaenopsis), *Cattleya labiata* (cattleya) and *Celosia cristata* (cockscomb). Flowers with special shapes are generally more expensive.

(4) Ornamental flower material

Ornamental flower materials refer to those with the shapes of dainty, dispersion or being gathered into a bundle, even with umbrella-shaped flowers, which are mainly used as filling flowers to decorate the whole profiling and work in with the main parts of flower arrangement. Common ornamental flowers include *Gypsophila Paniculata* (baby's breath), *Codariocalyx motoriu* (limonium), *Limonium Sinuatum* (myosotis), *Matricaria recutita* (chrysanthemum), *Tagetes patula* (peacockgrass), *Solidago decurrens* (solidago decurrens), *Asparagus setaceus* (asparagus), and *Asparagus cochinchinensis*(radix asparagus).

2.2 Conventional Herbal Cutting Flower Material

(1) Herbal cutting flower material for appreciating flowers (Table 2-1)

(2) Herbal cutting flower material for appreciating leaves (Table 2-2)

2.3 Conventional Woody Cutting Flower Material

(1) Woody Cutting Flower Material for Appreciating Branches and Flowers (Table 2-3)

(2) Woody Cutting Flower Material for Appreciating Leaves (Table 2-4)

(3) Woody Cutting Flower Material for Appreciating Fruits (Table 2-5)

表 2-1 常见草本观花花材及花语

商品名	中文名	科	拉丁学名	上市季节	花语及用法
大花葱	大花葱	百合科	Allium giganteum	夏	慈善、名誉；适用于各类插花
火鹤	花烛	天南星科	Anthurium spp.	全年	美好；可与文竹、肾蕨等组合
卡特兰	卡特兰	兰科	Cattleya hybrida	全年	敬爱、倾慕、高贵、优雅；为西方婚礼的高档用花
矢车菊	矢车菊	菊科	Centaurea cyanus	春末、夏	失恋、分离、忧伤
飞燕草	飞燕草	毛茛科	Consolida ajacis	春	慈悲的心；适用于各类插花
大丽花	大丽花	菊科	Dahlia pinnata	夏、秋	华丽、多情、优雅、尊贵
菊花	菊花	菊科	Dendranthema morifolium	全年	在中国象征长寿、高洁、坚强，欧美、日本等国黄、白菊搭配多用于祭祀
洋兰	石斛兰	兰科	Dendrobium nobile	全年	西方称之为"父亲之花"；适用于各类插花
康乃馨	香石竹	石竹科	Dianthus caryophyllus	全年	世界公认的"母亲之花"，象征温馨、慈祥
一品红	圣诞花	大戟科	Euphorbia pulcherrima	冬、春	圣诞祝福、热烈；适用于各类插花
洋桔梗	草原龙胆	龙胆科	Eustoma russellianum	春、秋	寓意真情；适用于各类插花
小苍兰	香雪兰	鸢尾科	Freesia refracta	冬、春	轻盈、秀丽；适用于各类插花
扶郎花	非洲菊	菊科	Gerbera jamesonii	全年	热烈、兴旺、发达，也含有"扶助郎君"之意
剑兰	唐菖蒲	鸢尾科	Gladiolus gandavensis	全年	步步高升、前程似锦
宿根霞草	满天星	石竹科	Gypsophila paniculata	全年	温馨；作填充花或调和色彩，表现虚景
向日葵	向日葵	菊科	Helianthus chinensis	春、秋	追随、仰慕、太阳的象征；适用于各类插花
黄金鸟	火鸟蕉	芭蕉科	Heliconia psittacorum	全年	高雅、孤傲、胜利、吉祥、高贵
萱草类	萱草	百合科	Hemerocallis fulva	夏	中国的"母亲花"
风信子	风信子	百合科	Hyacinthus orientalis	春	春讯、音信；宜做各种插花
爱丽丝	鸢尾	鸢尾科	Iris tectorum	春	象征圣母玛利亚；可作切花、切叶材料
火炬花	火炬花	百合科	Kniphofia uvaria	夏	希望、方向；可做各类插花
香豌豆	香豌豆	蝶形花科	Lathyrus odoratus	春、夏	轻盈、可爱
百合	百合类	百合科	Lilium spp.	全年	百年好合、纯洁、甜美；使用时需去掉花蕊
补血草	勿忘我	紫草科	Limonium sinensis	全年	永不忘记；多用于配花
杂种补血草	情人草	紫草科	Limonium spp.	全年	适用于各类插花；作填充花或调和色彩，表现虚景
紫罗兰	紫罗兰	十字花科	Matthiola incana	春、秋	永恒不变的爱；多用于西方式插花
荷花	荷花	睡莲科	Nelumbo nucifera	夏	为佛教圣花；象征纯洁和崇高，被誉为"水中芙蓉"

（续）

商品名	中文名	科	拉丁学名	上市季节	花语及用法
舞女兰	文心兰	兰科	Oncidium hybrida	全年	纯粹的爱
蝴蝶兰	蝴蝶兰类	兰科	Phalaenopsis spp.	全年	我爱你，幸福向你飞来
晚香玉	晚香玉	石蒜科	Polianthes tuberosa	冬、春	西方式插花应用较多
帝王花	山龙眼	山龙眼科	Protea spp.	冬	高贵、富庶
芹菜花	花毛茛	毛茛科	Ranunculus asiaticus	春	将获得财富；适宜各类插花
玫瑰	现代月季	蔷薇科	Rosa chinensis	全年	被誉为"爱情之花"，也可象征友谊、和平；适用于东、西方各类型插花
一枝黄花	黄莺	菊科	Solidago canadensis	全年	淡薄；多做配花
天堂鸟	鹤望兰	芭蕉科	Strelitzia reginae	全年	高雅、孤傲、胜利、吉祥、高贵
郁金香	郁金香	百合科	Tulipa gesneriana	春	博爱；适用于各类插花
马蹄莲	马蹄莲	天南星科	Zantedeschia aethiopica	全年	形如心，高贵、圣洁

表2-2 常见草本观叶花材

商品名	中文名	科	拉丁学名	上市季节	用法
番麻	龙舌兰	百合科	Agave americana	全年	多个排列使用，可形成很强的节奏感，可造"山景"
天冬草	天门冬	百合科	Asparagus cochinchinensis	全年	用于各类插花，多半作为打底
蓬莱松	蓬莱松	百合科	Asparagus myriocladus	全年	用于各类插花，既可作衬叶，也可作主枝
文竹	文竹	百合科	Asparagus setaceus	全年	用于各类插花，小型插花中可作主枝
一叶兰	蜘蛛抱蛋	百合科	Aspidistra elatior	全年	用于各类插花，并可撕、卷、折成各种艺术造型
黄椰子	散尾葵	棕榈科	Chrysalidocarpus lutescens	全年	适用于大、中型插花，并可修剪或编织各种造型
苏铁	苏铁	苏铁科	Cycas revoluta	全年	用于各类插花，并可制干涂上金银色
扇叶葵	蒲葵	棕榈科	Livistona chinensis	全年	一般修剪成各种形状后使用
龟背	龟背竹	天南星科	Monstera deliciosa	全年	适用于作背景叶或衬叶，也可代主枝
山苏叶	巢蕨	铁角蕨科	Neottopteris nidus	全年	用于各类插花
排草	肾蕨	骨碎补科	Nephrolepis auriculata	全年	用于各类插花
春羽	喜林芋	天南星科	Philodendron selloum	全年	与"龟背竹"同
虎尾兰	虎尾兰	百合科	Sansevieria trifasciata	全年	多用于背景叶或骨架花
棕榈	棕榈	棕榈科	Trachycarpus fortunei	全年	一般修剪成各种形状后使用

表 2-3 常见木本观花、观枝花材及花语

商品名	中文名	科	拉丁学名	上市季节	花语及用法
竹	竹	禾本科	Bambusoideae	全年	谦虚、坚强、有气节；多用于东方式插花及现代花艺，宜作主枝
木瓜花	贴梗海棠	蔷薇科	Chaenomeles speciosa	春	吉祥、富贵；多用于东方式插花
凤尾柏	'凤尾'柏	柏科	Chamaecyparis obtusa 'Filicoides'	全年	同松；宜作主枝或衬叶，也可作衬叶或造景
连翘花	连翘	木犀科	Forsythia suspensa	春	多用于东方式插花以及现代造型
水横枝、白婵	栀子	茜草科	Gardenia jasminoides	春	誉为"禅友"，为纯洁、贞节的象征
迎春	迎春	木犀科	Jasminum nudiflorum	早春	宜作弯曲造型，多用于东方式插花
白玉兰	白玉兰	木兰科	Magnolia denudata	春	纯洁、高贵、坚强；适用于各类插花
桂花	桂花	木犀科	Osmanthus fragrans	秋	团圆、富贵、秋天；宜作主枝或配花、衬叶
松	松	松科	Pinus spp.	全年	坚强、高洁、长寿；多用于东方式插花，宜作主枝及衬叶
梅花	梅	蔷薇科	Prunus mume	早春	坚贞、高洁、报春；多用于东方式插花
石榴	石榴	石榴科	Punica granatum	夏、秋	丰收、多子多孙；适用于各类插花，宜作倾斜主枝
柳	柳	杨柳科	Salix babylonica	春	春天；宜作主枝及附加枝
银芽柳	银芽柳	杨柳科	Salix leucopithecia	冬	多用于各类插花中线条的表现，宜作主枝及附加枝
龙柳	龙爪柳	杨柳科	Salix matsudana, var. matsaudana f. tortuosa	春	多用于各类插花中线条的表现，宜作主枝及附加枝
红瑞木	红瑞木	山茱萸科	Swida alba	初夏、秋	多用于插花中主枝及骨干枝

表 2-4 常见木本观叶花材及花语

商品名	中文名	科	拉丁学名	上市季节	花语及用法
米兰	米仔兰	楝科	Aglaia odorata	全年	适用于各类插花，也可作主枝
变叶木	变叶木类	大戟科	Codiaeum spp.	全年	变幻莫测、娇艳
熊掌叶	八角金盘	五加科	Fatsia japonica	全年	繁荣
橡皮树	橡皮树	桑科	Ficus elastica	全年	稳重、诚实、信任、万古长青、吉祥如意
茉莉花	茉莉花	木犀科	Jasminum sambaca	全年	忠贞、尊敬、清纯、贞洁、质朴、玲珑、迷人
九里香	九里香	芸香科	Murraya exotica	全年	爱情的俘虏

表 2-5 常见木本观果花材及花语

商品名	中文名	科	拉丁学名	上市季节	花语及用法
紫珠	白棠子树	马鞭草科	Callicarpa dichotoma	冬	果实圆润
柿	柿	柿树科	Diospyros kaki	冬	自然美
老虎刺、鸟不宿	枸骨	冬青科	Ilex cornuta	冬	生命
广玉兰	广玉兰	木兰科	Magnolia grandiflora	秋	多用于东方式插花
石榴	石榴	石榴科	Punica granatum	夏、秋	丰收、多孙子；可用于各类插花
火棘	火棘	蔷薇科	Pyracantha fortuneana	冬	满堂红，在新的一年里步步高升
狐脸茄	乳茄	茄科	Solanum mammosum	冬	五福临门、金玉满堂、富贵发财、子孙繁衍不息、代代相传

Chapter 2 Flower Materials and Conventional Equipment for Flower Arrangement

Table 2-1 Conventional herbal cutting flower material for appreciating flowers

Trade Name	Engish Name	Family Name	Scientific Name	Time to Market	Floral Language and Usage
allium giganteum	allium giganteum	Liliaceae	Allium giganteum	summer	charity, reputation, suitable for all types of flower arrangement
anthurium andraeanum	anthurium	Araceae	Anthurium spp.	whole year	a symbol of beauty, can be combined with asparagus fern, fern
maurizio cattelan	maurizio cattelan	Orchidaceae	Cattleya hybrida	whole year	noble, elegant, as the up market flowers in western weeding
bachelor's buttons	bachelor's buttons	Asteraceae	Centaurea cyanus	later spring, summer	breaking up, grief, separation
delphinium	delphinium	Ranunculaceae	Consolida ajacis	spring	kind-heart, adaptable to all kinds of flowers
dahlia	dahlia	Asteraceae	Dahlia pinnata	summea utumn	gorgeous, elegant, noble, amorous
chrysanthemum	flovists derdranthema	Asteraceae	Dendranthema morifolium	whole year	in chinese tradition, as the symbol of longevity, nobility, strong; in europe and the united states, japan, using yellow chrysanthemum for sacrifice
cattleya	dendrobium	Orchidaceae	Dendrobium nobile	whole year	known as "the flower of father" in western, suitable for all types of flower arrangement
dianthus caryophyllu	carnation	Caryophyllaceae	Dianthus caryophyllus	whole year	well known internationally "flower of mother", symbolize warmth, kindness
poinsettia	christmas flower	Euphorbiaceae	Euphorbia pulcherrima	winter, spring	christmas greetings, warm, suitable for all types of flower arrangement
eustoma grandiflorum	eustoma russellianum	Gentianaceae	Eustoma russellianum	Spring, autumn	true love, suitable for all types of flower arrangement
freesia	freesia	Iridaceae	Freesia refracta	winter, spring	lithe, beautiful, suitable for all types of flower arrangement
gerbera	barberton daisy	Asteraceae	Gerbera jamesonii	whole year	a symbol of warmth, prosperous, wealthy, also means "support the husband"
gladiolus	gladiolus gandavensis	Iridaceae	Gladiolus gandavensis	whole year	be promoted step by step, a bright future
gypsophila poniculata	baby's breath	Caryophyllaceae	Gypsophila paniculata	whole year	warm, being the filler flower or counterbalancing the color, performing the "ritual" vision
sunflower	sunflower	Asteraceae	Helianthus chinensis	spring, autumn	a symbol of following up the sun, suitable for all types of flower arrangement
heliconia	heliconia bihai	Musaceae	Heliconia psittacorum	whole year	a symbol of elegance, aloof, victory, auspicious, noble
hemerocallis	hemerocallis	Liliaceae	Hemerocallis fulva	summer	chinese "flowers of mother"
hyacinthus orientalis	hyacinth	Liliaceae	Hyacinthus orientalis	spring	spring signor, good news, suitable for all types of flower arrangement
iris	allis	Iridaceae	Iris tectorum	spring	a symbol of santa maria, can be cut as leaf material
kniphofia uvaria	kniphofia uvaria	Liliaceae	Kniphofia uvaria	summer	hope, direction, suitable for all types of flower arrangement
sweet pea	sweet pea	Papilionaceae	Lathyrus odoratus	spring, summer	light, sweet

(continued)

Trade Name	English Name	Family Name	Scientific Name	Time to Market	Floral Language and Usage
lily	lilium	Liliaceae	*Lilium* spp.	whole year	true love which will last a hundred years, pure, sweet, should remove the stigma when using
limonium sinense	forget-me-not	Boraginaceae	*Limonium sinensis*	whole year	never forget, for setting off other flowers
hybrid chinese sealavender	desmodium gyrans	Boraginaceae	*Limonium* spp.	whole year	suitable for all types of flower arrangement, to be the filled flower or counterbalance the color, to perform the "irtual" vision
violet	violet	Cruciferae	*Matthiola incana*	spring, autumn	eternal love, for western style flower arranging
lotus	lotus flower	Nymphaeaceae	*Nelumbo nucifera*	summer	being the buddhist holy flowers, a symbol of purity and loftiness, being known as the "water lotus"
oncidium ampliotum	dancing-douorchid	Orchidaceae	*Oncidium hybrida*	whole year	represents pure love
phalaenopsis	phalenopsis	Orchidaceae	*Phalaenopsis* spp.	whole year	i love you, wish you a happy life
tuberose	tuberose	Amaryllidaceae	*Polianthes tuberosa*	winter, spring	implicated more in western style flower arranging
protea	protea	Proteaceae	*Protea* spp.	winter	noble, wealthy
ranunculus asiaticus	ranunculus	Ranunculaceae	*Ranunculus asiaticus*	spring	wealthy, adaptable to all kinds of flowers
rose	chinese rose	Rosaceae	*Rosa chinensis*	whole year	internationally known as "the flower of love", also on behalf of friendship, peace, used both in the eastern and western flower arrangement
solidago virgo-aurea	songbirds	Asteraceae	*Solidago canadensis*	whole year	noble, for setting off other flowers
paradiesvogel	bird of paradise	Musaceae	*Strelitzia reginae*	whole year	a symbol of elegance, aloof, victory, auspicious, noble
tulips	tulips	Liliaceae	*Tulipa gesneriana*	spring	fraternity, suitable for all types of flower arrangement
calla	calla	Araceae	*Zantedeschia aethiopica*	whole year	shaped like a heart, noble, holy

Table 2-2 Conventional herbal cutting flower material for appreciating leaves

Trade Name	English Name	Family Name	Scientific Name	Time to Market	Usage
american aloe	century plant	Liliaceae	*Agave americana*	whole year	used as a plurality of arrangement, can form a strong sense of rhythm, can be built "mountain vision"
asparagus	asparagus	Liliaceae	*Asparagus cochinchinensis*	whole year	suitable for all types of flower arrangement, probably as a bottoming
asparagus myrioeladus	asparagus myrioeladus	Liliaceae	*Asparagus myriocladus*	whole year	suitable for all types of flower arrangement, not only can be used as the greens, also as the bough
setose asparagus	asparagus fern	Liliaceae	*Asparagus setaceus*	whole year	suitable for all types of flower arrangement, as main branch in small works
aspidlsiraelatior	aspidistra	Liliaceae	*Aspidistra elatior*	whole year	suitable for all types of flower arrangement, used as tearing, rolling, folding and various art forms

Chapter 2 Flower Materials and Conventional Equipment for Flower Arrangement

(continued)

Trade Name	English Name	Family Name	Scientific Name	Time to Market	Usage
chrysalidocarpus lutescens	bamboo palm	Arecaceae	Chrysalidocarpus lutescens	whole year	suitable for large, medium-sized floral arrangement, and may be trimmed or woven into various shapes
cycad	cycad	Cycadaceae	Cycas revoluta	whole year	suitable for all types of flower arrangement, can be dried and painted with gold and silver
fan palm	sabal	Arecaceae	Livistona chinensis	whole year	cut into various shapes and then be used
turtle	turtle	Araceae	Monstera deliciosa	whole year	applicable to be both background leaves or greens and bough
neottopteris nidus	neottopteris nidus	Aspleniaceae	Neottopteris nidus	whole year	suitable for all types of flower arrangement
lysimachia sikokiana	tuber fern	Davalliaceae	Nephrolepis auriculata	whole year	suitable for all types of flower arrangement
philodendron selloum c.koch	philodendron	Araceae	Philodendron selloum	whole year	same as monstera
snake plant	snake plant	Liliaceae	Sansevieria trifasciata	whole year	for background leaves or skeleton flowers
palm	palm	Arecaceae	Trachycarpus fortunei	whole year	cut into various shapes and then be used

Table 2-3 Conventional woody cutting flower material for appreciating branches and flowers

Trade Name	English Name	Family Name	Scientific Name	Time to Market	Floral Language and Usage
bamboo pole	bamboo	Gramineae	Bambusoideae	whole year	modesty, strong, has integrity, for oriental flower arranging and modern style, suitable for main branch
japanese quince	chaenomeles speciosa	Rosaceae	Chaenomeles speciosa	spring	lucky, wealthy, for oriental flower arranging
chamaecyparis obtusa cv	chamaecyparis obtusa cv	Cupressaceae	Chamaecyparis obtusa 'Filicoides'	whole year	same as pine, to be the main branch or greens or build "vision"
forsythia viridissima	forsythia	Oleaceae	Forsythia suspensa	spring	for oriental flower arranging and modern design
gardenia augusta	water bough	Rubiaceae	Gardenia jasminoides	spring	known as the friend of "chan", as a symbol of purity, chastity
winter jasmine	winter jasmine	Oleaceae	Jasminum nudiflorum	early spring	suitable for curved shapes, for oriental flower arranging
magnolia	magnolia	Magnoliaceae	Magnolia denudata	spring	pure, noble, strong, suitable for all types of flower arrangement
osmanthus fragrans	osmanthus fragrans	Oleaceae	Osmanthus fragrans	autumn	reunion, rich, autumn, suitable for the main branch or greens
matsueda	pine	Pinaceae	Pinus spp.	whole year	strong, noble, longevity, for oriental flower arranging, to be the main branch or greens
plum blossom	plum blossom	Rosaceae	Prunus mume	early spring	stout, noble, signor of spring, for oriental flower arranging
pomegranate	pomegranate	Punicaceae	Punica granatum	summer, autumn	harvest, having many children, suitable for all types of flower arrangement, prefer to be inclined branch

Trade Name	English Name	Family Name	Scientific Name	Time to Market	Floral Language and Usage
willow twig	willow	Salicaceae	*Salix babylonica*	spring	used for the spring theme and the changeable curve and advised to make additional branches
salix gracilistyla	salix gracilistyla	Salicaceae	*Salix leucopithecia*	winter	often used to highlight the curves, prefer to be the main branch and additional branch
salix matsudana 'tortusoa'	salix matsudana 'Tortuosa'	Salicaceae	*Salix matsudana* var.*matsaudana* f. *tortuosa*	spring	often used to highlight the curves, prefer to be the main branch and additional branch
swida alba	swida alba	Cornaceae	*Swida alba*	early summer, autumn	for the main branches and the backbone branches

Table 2-4 Conventional woody cutting flower material for appreciating leaves

Trade Name	English Name	Family Name	Scientific Name	Time to Market	Floral Language and Usage
milan	milan	Meliaceae	*Aglaia odorata*	whole year	suitable for all types of flower arrangement, can be the main part
change Leave wood	change leave wood	Euphorbiaceae	*Codiaeum* spp.	whole year	change unpredictably, chameleon, delicate and charming
fatsia japonica	fatsia japonica	Cornaceae	*Fatsia japonica*	whole year	boom
ficus elastica	ficus elastica	Moraceae	*Ficus elastica*	whole year	stable, honest, trust, a symbol of longevity, lucky
jasmin	jasmin	Oleaceae	*Jasminum sambac*	whole year	loyal, respectful, pure, simple, exquisite, charming
common jasminorange	common jasminorange	Rutaceae	*Murraya exotica*	whole year	prisoner of love

Table 2-5 Conventional woody cutting flower material for appreciating fruits

Trade Name	English Name	Family Name	Scientific Name	Time to Market	Floral Language and Usage
beautyberry	callicarpa dichotoma	Verbenaceae	*Callicarpa dichotoma*	winter	mellow fruit
persimmon	persimmon	Ebenaceae	*Diospyros kaki*	winter	natural beauty
ilex cornuta	pterolobium punctatum	Aquifoliaceae	*Ilex cornuta*	winter	vitality
magnolia grandiflora	magnolia grandiflora	Magnoliaceae	*Magnolia grandiflora*	autumn	for oriental flower arranging
pomegranate	pomegranate	Punicaceae	*Punica granatum*	summer, autumn	harvest, having many children, suitable for all types of flower arrangement
huo zao	huo zao	Rosaceae	*Pyracantha fortuneana*	winter	be promoted step by step in the new year
fox face eggplant	Abiu	Solanaceae	*Solanum mammosum*	winter	a symbol of happiness, wealth, because of the five generations under one roof, which means the descendants multiply without rest, be handed down from age to age

2.4 常用插花器具

2.4.1 常用插花器皿

（1）传统东方式插花器皿

在传统东方式插花中，花器是其构成的一个重要部分。传统东方式插花的器皿有：瓶、盘、缸、碗、筒、篮六大类型。材质有陶、瓷、玻璃、水晶、玉、铜、银、漆器、竹木等。

①瓶花：瓶花为中国插花的起源。佛教认为瓶可盛装万物，在汉语中"瓶""平"同音，因此瓶插花又有吉祥、平安之意。瓶花形体高昂，象征宗庙、高堂或崇山峻岭，所以理念花多以瓶插。常见的花瓶口径较窄较小，瓶肚较大，瓶体较高，盛水量较大，所需花材较少。

②盘花：盘状花器的特点是口径大、储水浅，一般用于短花枝直接浸润或浮于浅水的插花。古人常以"盘"为大地，用花材插于其内，表现某些自然景观。

③缸花：缸的形状与今日的钵颇似，适宜插丰盛的花材，表现块体。

④碗花：东方式三主枝插法中的盆式插花即源于碗花。碗花挺直，不偏不倚。

⑤筒花：利用自然界竹子中空有节的特点，锯断或打洞盛水插花。筒花是东方插花最自然的花器，一直沿用至今。筒花还可以挂于壁上。

2.4 Conventional Equipment for Flower Arrangement

2.4.1 Conventional Equipment for Flower Arrangement

(1) Traditional Eastern Utensil for Flower Arrangement

The use of equipment is an important component in traditional flower arrangement in some eastern countries. There are mainly six types of traditional oriental receptacles: vase, plate, crock, bowl, bucket and basket. In materials it can be divided into pottery, porcelain, glass, crystal, jade, copper, silver, lacquer, bamboo, wood, etc.

① Flowers in vase: Flowers in vase are the origin of Chinese flower arrangement. Buddhism believes that vase can contain everything; in Chinese, "vase" has the same pronunciation with "safety" which means lucky and safe. Mostly, the shape of vase is usually made tall and big, symbolizing the ancestral temple, a hall with high ceiling and lofty hills. So, most of conceptive flower arrangement are created with vase. Common vase has narrow caliber, big belly and high body so that it can contain much water but need less floral materials.

② Flowers in plate: Flowers in plate has the feature that caliber is big and the water is shallow, so it is generally used for short branches, infiltrating or floating directly in shallow water. In ancient times, "plate" used to be regarded as the "earth", so that inserting flowers into it was to show natural landscape.

③ Flowers in crock: The shape of crock is similar to the earthen bowl today, which is suitable to insert abundant flower materials to represent block structure.

④ Flowers in bowl: Flowers in bowl: they are potted flowers from which the oriental flower arrangement of three major branches has derived. Flowers in bowl are straight without leaning to any side.

⑤ Flowers in tube: In making it, we arrange flowers by sawing off or making holes to pour water in bamboo, which

⑥篮花：篮状花器一般由柳、竹、草、藤编制，外形如篮，插花成品称为篮花。

（2）传统西方式插花器皿

传统西方式插花器皿使用的初衷是盛水，以保持花的新鲜度。只要能满足插花的必要条件，任何容器都可以用来插花。一般为敞口，以利于盛放较多的草本花材。

（3）现代插花器皿

现代插花作品中，对器皿的选择不拘一格，各种质地、形状、颜色的器皿都可应用，有些日常生活中的容器和用品也可用于插花。

2.4.2 常用工具和辅件

（1）工具

①必备工具：

剪刀：主要用来剪切花枝、修剪叶片等，由于各种花材的质地及硬度不同，最好准备一把质量好的枝剪，用来剪硬枝条；一把普通剪刀，用来修剪草本花枝和叶片。最好能准备不同型号的剪刀，以备不时之用。

小刀：主要用来切割花泥，而质量好的利刀可用来修剪和切削花枝，以及雕刻和去皮等。刀的切面较平滑，不易挤压花茎，对花材吸水保鲜更有利。

锯：主要用于修剪加工较硬、粗的木本花材。

is hollow inside and has knots on the surface. It is the most natural container in the east and still in use today. Flowers in tube can also be hung on the wall.

⑥ Flowers in basket: Flowers in basket: Basket-type flower containers are generally made of willow, bamboo, grass and vine. They are shaped like baskets and "flowers in basket" is the name for finished arranged flowers.

(2) Traditional Containers in Western Flower Arrangement

The original purpose for using traditional western flower containers is to hold water to maintain the freshness of flowers. Any type of containers can be used to arrange flowers as long as they meet the requirements. Generally, these containers are open which are favorable to hold more herbaceous flowers.

(3) Modern Flower Containers

For the products of the modern flower arrangement, a diverse set of containers with different textures, shapes and colors can be chosen to arrange flowers. Some daily utensils and other goods can also be used for holding arranged flowers.

2.4.2 Conventional Tools and Their Accessories

(1) Tools

① Essential tools:

Scissors: They are mainly used for cutting branches and trimming leaves. On account of the difference of flower texture and hardness, it is better to prepare a pair of scissors with good quality to cut hard branch and a pair of ordinary scissors to trim herbaceous flower branches and leaves. We'd better be equipped with scissors of different types for unexpected needs.

Knives: They are mainly used for cutting flower mud, while sharp knives of superior quality can be used to trim and cut flower branches, engrave as well as peel. The smooth surface of the knife makes it less likely to squeeze stems so that flower can better suck water and stay fresh.

Saws: They are mainly used for cutting and processing harder and thicker arborescent flowers.

Wrenches or diagonal cutting pliers: Ordinary scissors

大力钳或斜口钳：由于一般人造花花梗均用铁丝或钢丝外加塑料或纸张做成，一般剪刀及枝剪都剪不动，即使勉强剪动或剪断，对剪刀的刀口伤害很大，因此需要使用大力钳或斜口钳来剪。

花泥：花泥由酚醛塑料发泡制成，使用方便，但插后空洞不能复原，使用1~2次就会散裂报废。花泥分为鲜花花泥和干花花泥两种。

鲜花花泥：中国港台地区称为花泉。绿色，有多种形状，常见为长方体，长23cm，宽11cm，高7cm。干时极轻，吸水后很重，保水性能好，一般10d后保水率仍在90%。目前市场上的花泥分为标准型和加强型，标准型适合日常普通插花用，各类细软花枝及多数鲜切花均可使用；加强型即增加强度的花泥，适合大型插花使用，尤其适合粗大花枝使用，具有支撑力强和稳定性好的特点。鲜花花泥使用前应提前将花泥平放在水中使其自然吸水沉入水中，一般经1~3min即可完全吸水；忌贪图快而将花泥全部摁入水中，造成花泥中部干心。干花花泥也叫泡棉，是人造花、干燥花使用最广泛的固定材料。大小跟鲜花花泥相同，质地稍硬，忌水，应在干燥环境中使用。由于质轻，最好用热熔胶在花器做合理的固定。

are not sharp enough to cut the peduncles of artificial flowers because they are generally made of iron wires, steel wires, plastics or papers. A forced cut will intensify the blade damage, so it is necessary to use wrenches or diagonal cutting pliers.

Flower mud: Flower mud is made from phenolic foams. It is convenient to use but holes on the mud cannot be recovered after inserting. When used once or twice, the mud will crack and be out of service. Flower mud consists of fresh flower mud and dried flower mud.

Flower mud: It is called huaquan (flower spring) in Hong Kong and Taiwan areas of China. Such kind of mud is usually green and has various shapes, of which cuboid is the most common, being 23cm long, 11cm wide and 7cm high. Dried flower mud is extremely light but after water absorption it is quite heavy and stays wet for a long time. Generally, water-retention rate of it will still be 90% after 10 days, indicating a unique property of holding water. Standard and fortified flower mud is currently supplied in market. The standard ones are suitable for ordinary daily flower arrangements and applicable to all kinds of soft flowers and most of fresh cut flowers. Fortified ones refer to the flower mud with additional strength. They are designed to arrange large flowers, especially thick flowers with strong support and good stability. Before being used, fresh flower mud should be placed flat in water before letting them naturally sink into the bottom by absorbing water. In general, it takes one to three minutes to absorb enough water. You shouldn't press the whole mud into water just in pursuit of speed, or the mud center will remain dry. Foam is another name for dried flower mud, which is the most widely used material to fix artificial flowers and dried flowers. Dried flower mud is in the same size as fresh flower mud, but have relatively hard texture and should be used in the dry environment without water. It is suggested that hot melt adhesives be used to appropriately fix them in flower container.

②其他工具：

水桶及花筒：水桶主要用于清洗和储养花材。一般使用有提梁的大号和中号的塑料桶即可。花筒一般放在花架上，用来插摆花材样品。有塑料、玻璃、陶瓷等多种，塑料筒最方便、轻巧、经济；陶筒最有利于花材保鲜。

花插：花插也称剑山，是东方传统插花器具，一般用于水盆和盘等浅口插花容器的主要花枝的固定。花插系由铅锡底座和铜或不锈钢针头浇铸而成，利用金属的重量和钉子扎花能力，达到固定花枝的作用。具圆形、方形、棱形、半月形等。规格有大、中、小多种。花插常见有用金属制成的古钱状花插和用玻璃制成的蜂巢状花插。

（2）辅件

①一般辅件：

铁丝：主要用于花枝的整形加固或弱枝及变形花朵的绑扎处理，如对花材进行缠绕、绑扎、支撑、弯曲、造型等。较多使用18~28号铁丝，号码越大，铁丝越细。最好用绿棉纸或绿漆做表面处理。根据花材软硬，采用不同型号的铁丝：木本花材及较粗大的叶材一般使用18~20号；较软的草本花材及胸花、头花用22~24号。

棉纸贴布：指不同颜色的特殊皱纹纸间加入黏胶制成的包装带。以绿色居多，故又称绿胶带或棉纸胶带。棉纸贴布常用来包裹铁丝，掩盖铁丝加工的痕迹。使用棉纸时应稍稍拉伸缠绕，使皱纹间的黏胶露出来，以便牢固地裹紧花材。

② Other tools:

Bucket and flower tube: Bucket is mainly used for cleaning and holding flower materials. Large and medium plastic buckets with lifting handles are usually used. Flower tube is generally displayed on flower shelves for arranging flowers. You can see plastic, glass and ceramic tubes, of which plastic tube with a light weight is the most convenient, handy and cost-effective while ceramic tube is most conducive to maintaining flower freshness.

Flower receptacle: As a traditional tool in oriental flower arrangement, flower receptacle is also known as Jianshan (sword mountain). It is commonly used as the tool for fixing flowers in such shallow containers as basins and plates. Flower receptacle, which is cast from lead-tin pedestals and needles made from copper or stainless steel, function as fixing flowers by taking advantage of the weight of metals and the fixing property of nails. There are shapes of being round, square, rhombic or half-moon-shaped, with large, medium and small specifications. It is common to see ancient-coin-shaped receptacles made from metal and honeycomb ones made from glass.

(2) Accessories

① General accessories:

Iron wire: It is mainly used for reshaping flowers or binding weak branches and deformed flowers, including twining, bundling, backing, bending and shaping. Iron wires of Size 18 to 28 are used more frequently. The larger the size, the thinner the wire. Green cotton paper or green paint are better materials for covering the surface. Selection of wires depends on the hardness of flowers: wires of Size 18 to 28 are used for arborescent flowers and thick branches while those of Size 18 to 28 are for soft herbaceous flowers, flower brooches and headdress flowers.

Cotton paper patch: It is packaging tapes made from special crepe paper with different colors and glued to each other through mucilage glue. Since most of them are green, they are often called green tape or cotton paper tape. Cotton paper is used to wrap wire in order to cover the processing

玫瑰去刺钳：主要用于玫瑰、月季去刺和其他需要去叶的花材，如一枝黄花、香石竹、百合等。

订书机：在插花造型的技术处理中，可用于卷叶造型固定，如巴西铁、剑叶龙血树、一叶兰等，可用订书机固定。

透明胶带：可以用于花材的特殊要求处理，也可用来固定花泥、包装纸等。

竹签：插入柔软易倒伏的花梗内，防止其倒伏。

喷水壶：花材整理修剪后，未插之前及插好后都要喷水，以保持花材新鲜。

花胶：主要用于粘贴花材。分为冷胶和热胶，冷胶一般用于粘贴鲜切花材料，热胶多用于干花插花和人造花插花。

②装饰辅件：装饰辅件一般指在插花作品周围摆放的一些小摆设物，以增添气氛。辅件的大小、形态和摆设位置必须与插花作品相衬，不能喧宾夺主。常用的装饰辅件如下：

垫座：是指铺垫在插花作品下面的一种用具，其作用是烘托作品、增添美感。垫座除名贵的花梨木、红木制成的几架外，还有树根制品、木板、竹垫、草垫、塑胶垫等。常应用于我国古典插花中。

traces of it. When using cotton paper, you should slightly stretch and twine it to expose the glue so that the flowers can be firmly wrapped.

Plier without thorn: It is mainly used for cutting the thorns of *Rosa rugosa*, *R.chinensis* and other flowers that need to cut leaves, such as *Solidago decurrens*, *dianthus caryophyllus* and *Lilium* spp.

Stapler: It is used to fix the shape of curly leaves in the technical process of flower arrangement. For instance, *Dracaena fragrans*, *D. cochinchinensis*, and *Aspidistra elatior* can be used in this way.

Transparent tape: It meets the special requirements of flower arrangement and can also be used to fix flower mud and packaging paper.

Bamboo skewer: Insert it into soft peduncle that is prone to toppling so as to prevent this from clapping.

Watering can: Flower should be watered before and after arrangement and after trimming so as to keep flowers fresh.

Flower glue: It is mainly used for stick flowers together. Flower glue can be divided into cold glues and hot glues, with the former for fresh cut flowers and the latter for dried flowers and artificial flowers.

② Decorative accessories: Decoration accessories generally refer to the small ornaments placed around the arranged flowers, aiming at creating a pleasant atmosphere. These accessories should be in line with rather than overshadow the arranged flowers in size, shape and placement. The commonly used decoration accessories are as follows:

Pedestal: Pedestal is what is placed under the arranged flowers so as to foil flowers and enhance beauty. Apart from rare rosewood and mahogany stands, there are also tree-root-made articles, woods, bamboo mats, grass mats and plastic mats. They are often used in classical flower arrangement in China.

Decorative packaging ribbons: Strings include ribbon, and the ribbons of plastic PP, cloth, yarn, mesh belts, non-

包装装饰绳带：带类包括缎带、塑料制PP带、布带、纱带、网带、无纺布带、纸带、尼龙带、丝带、麻带等。绳类常见用于绑扎的麻绳、尼龙绳等，装饰用的有纸绳、绞绳、拉菲草等。

包装装饰面料：包括纸制品、无纺布、纱和网纱等。纸制品有手揉纸、金花纸、瓦楞纸、皱纹纸、皱卷纸等。无纺布又称不织布，有平板状、卷筒状。纱一般有珍珠纱。网纱有婚纱、人造纱等。

装饰辅料：常见有心形棒、金粉、天使发丝、珠子、珠链、花篮卡、艺术铝线、魔术丝、插卡棒等。

woven belts, paper tapes, nylon tapes, silk ribbons and hemp belts. Hemp rope and nylon rope are commonly used for binding while paper rope, twisted rope and raffia are for decorative purposes.

Decorative packaging materials: They include paper-made products, non-woven fabrics, yarn and mesh yarn. Paper-made products are made up of hand-rubbing paper, golden flower paper, corrugated paper, crepe paper and wrinkled paper. Non-woven fabrics are those of fabrics that have not been woven, such as tabulate fabrics and scroll-like fabrics. Pearl yarn is common in yarn and mesh yarn as wedding dresses and synthetic yarns.

Decorative accessories: Common decorative accessories include heart-shaped rod, gold powder, angel silky, bead, bead chain, flower basket card, artistic aluminum yarn, magic silk and card stick.

第 3 章　礼仪用花

Chapter 3　Flower Arrangement for Etiquette

3.1 婚庆用花

3.1.1 新娘花

（1）捧花

新娘手捧花使用的主要花材有切花月季、百合、兰花，以及一些小花和绿叶。国内的新娘手捧花一般选择19，24枝等吉利数字的切花月季加2枝百合、满天星组合而成，切花月季的数量也可根据新娘的年龄来定。

新娘捧花可以制作成圆形、瀑布形、三角形、新月形、S形等。一般用捧花托插制（图3-1），也有用缎带包装的（图3-2）。

3.1 Wedding Flower Arrangement

3.1.1 Decorative Flower for Bride

(1) Bouquet

The main floral materials of bride bouquet include cut *Rosa chinensis*, *Lilium* spp. var. *viridulum*, *Cymbidium* spp., some small flowers and green leaves. Chinese bride bouquet generally consists of lucky numbers such as 19 or 24 branches of cut *Rosa chinensis* and 2 branches of *Liliam* spp. and Gypsophila paniculata. The number of cut *Rosa chinensis* can be determined according to the age of the bride.

Bride bouquet can be made into the shapes of round, waterfall, triangle, crescent, S and etc. It is generally inserted with holding receptacle, (Figure 3-1) , or wrapped with paper packaging (Figure 3-2).

图3-1 捧花托
Figure 3-1　Holding receptacle

图3-2 新娘捧花
Figure 3-2　Bride bouquet

设计新娘手捧花应考虑新娘所穿礼服的颜色、式样及新娘的发型、肤色、体形、婚礼的整体气氛等因素。新娘所穿礼服下摆如果是直筒形或者A字形，可以选择瀑布式下垂捧花；如果新娘所穿礼服下摆较大，腰身紧缩，应选择圆形捧花。西式捧花一般颜色淡雅，白色、粉色居多，形式也不固定；而中式捧花则色彩艳丽、鲜艳。至于选择哪种形式的捧花，应依据个人喜好和婚礼气氛而定。如果婚礼是中式的，应选择中式捧花；如果婚礼是西式的，应选择色彩淡雅的西式捧花。此外，满天星、情人草、文竹、勿忘我、郁金香、蝴蝶兰、花烛、马蹄莲、常春藤等花也常常在新娘捧花中使用。

（2）头花

新娘的头花非常重要，能够创造出喜庆的氛围，衬托出新娘的娇美。头花制作应选择易配合发型的花材，头花的体量要与新娘的整体造型相适应。结婚是人一生中最隆重的大事，很多新娘会选择盘头。盘头时可把鲜花插在盘好的头发中。一般插在盘好的发堆处，花朵下面衬一些绿叶，会起到画龙点睛的作用。如果感觉头花插得不牢固，可再用黑卡子固定，插好的头花只露花头，不可看见花枝，如图3-3所示。

头花的设计应考虑与新娘的肤色、脸形、身高相配。头花的设计不可过大，如果脸形过小则慎用大朵百合，一般会用石斛兰、玫瑰等。需要注意的是，新娘的头花、胸花、手捧花所使用的花材色彩应配套，互相呼应。

The color and style of the bride's dress, the bride's hair style, skin color, figure and the whole atmosphere of the wedding should be taken into consideration when designing bride bouquet. If the hem of the bride's dress is straight cylinder or A glyph, we can choose the shape of waterfall; If the bride's dress has a big lower hem and tightening waist, we should choose round bouquet. Western-style bouquet are generally light in color, mostly white and pink, whose forms are not fixed. However, the Chinese-style bouquet are colorful and bright. As to choosing what kind of flowers, it can be decided by individual preference and wedding atmosphere. If the wedding is Chinese-style, we should choose Chinese-style bouquet; if it is a western-style wedding, we should choose western-style bouquets which have quietly elegant color. In addition, *Gypsophila paniculata*, *Codariocalyx motorius*, *Asparagus setaceus*, *Myosotis silvatica*, *Tulipa gesneriana*, *Phalaenopsis aphrodite*, *Anthurium* spp., *Zantedeschia aethiopica*, *Hedera nepalensis* var. *sinensis* (ivy) and other flowers are often used in the bride bouquet.

(2) Headdress Flower

The bride's headdress flower is very important, which can create a festive atmosphere and foil the beauty of bride. Materials easy to match the shape of head should be chosen when making headdress flower. The size of headdress flower should be compatible with the bride's overall figure. Wedding is the grandest event in people's life so a lot of brides will choose to roll up their hair. When hair being rolled up, fresh flowers can be inserted into the hair. They are often inserted in the stack of hair with some green leaves as a finishing touch. If the flower is not secured, a black clip can be used to fix. Headdress flower just shows their heads not branches, as shown in Figure 3-3.

The design of headdress flower should match the bride's skin color, face and height. Headdress flower shouldn't be too large. If the face is too small, large *Lilium* spp. should not be used while *Dendrobium nobile* (dendrobium) and *Rosa rugosa* are preferable. One thing to note is that the material and color used in bride's headdress flower, brooches and hand bouquet should be matching.

图3-3 新娘头花
Figure 3-3 Bride's headdress flower

（3）腕花

腕花是新娘佩戴在手腕上的花。婚礼上新人会在胸前佩戴胸花，而胸花后的挂针会损坏礼服，新人也开始用国外的腕花代替胸花，腕花渐渐流行起来。腕花一般由两朵主花和辅材构成，主花材要和婚礼用花一致。一般在迎亲的时候新郎为新娘把腕花佩戴在左手或者右手，多数是佩戴在左手，与佩戴婚戒的手一致。因为手腕花为人体花饰的一种，所以选择材料的时候应该注意以下几点：无毒，无刺激性，无尖锐刺等。一般用28#铁丝制作；比较大的鲜花用26#铁丝处理。

3.1.2 花车

接亲的花车由车头花、车门花、镶边花三部分组成，高档花车还可以装饰车顶和车尾，如图3-4所示。

(3) Wrist Flower

Wrist flower is for the bride to wear on her wrist. The couple will wear a flower on chest in the wedding, but the brooch behind the flower can damage the dress, so the couple begins to use foreign wrist flower instead and wrist flower has also gradually become popular. Wrist flower generally consists of two main flowers and auxiliary materials. The main materials must be fit to the wedding flowers. Generally when the bridegroom comes to escort the bride to the wedding, he will help her wear the wrist flower on either of her hand, mostly the left hand, being consistent with the hand of wearing a wedding ring. Because the wrist flower is one of the body decoration, material selection should be paid attention to: non-toxic, without excitant, without sharp thorn, and etc. Usually it is made of iron wire No. 28 and larger flower of iron wire No. 26.

3.1.2 Floats/Festooned Vehicle

Flowers used to decorate the floats for picking up bride consist of three parts, namely front flower, door flower and embroidered flower. High-grade floats can be also decorated at the roof and the rear, as shown in Figure 3-4.

图3-4 花车
Figure 3-4 Float

（1）车头花

在制作车头花时，可以把三角形或圆形、椭圆形插花插在花车花泥上，再用吸盘固定。最常规的做法是，首先在带吸盘的花车花泥上面绑几个小吸盘，增大其吸力，以便车头花更稳固地固定在车头上，如图3-5所示。

然后，在花车吸盘上面插入配叶，用切花月季等主花材插出扇面的形状，花与花之间保持一指距离。同时，用百合花定出车头花的高度。每插入一层切花月季，在切花月季上面插入一层满天星，再在满天星上面插入一层切花月季，每层花的高度比上一层矮一个花头，这样一层层地插入。最后，插入配叶盖住花泥，再将不规则的地方修整成形。在新娘花车的车头，除了放置车头插花以外，还可以放置红色缎球、纱球、一对小人等，如图3-6所示。注意用塑料丝带穿过缎球系一下，以方便粘贴在汽车上面。

(1) Front flower

Front flowers are made by putting triangular or circular, oval flowers in the flower mud and being fixed in the front with suckers. The routine approach is to tie several small suckers to the flower mud of the floats firstly to increase its suction, so that front flowers can be firmly fixed on the front, as shown in Figure 3-5.

Then, insert matching leaves in the plate and shape sector with the main materials such as cut *Rosa chinensis*. There is one-finger distance between flowers at the same time. Use *Lilium* spp. to determine the height of the front flowers. Each layer of cut *Rosa chinensis* should be inserted into a layer of *Gypsophila paniculata* and then put another layer of cut *Rosa chinensis* on it. The height of each layer of flowers is about a flower head shorter than another. Insert layers by layers. Finally, insert leaves to cover flower mud, and then reshape irregular places. On the front of the bride's float, besides front flower arrangement, red satin balls, gauze balls and a pair of little couple dolls can also be placed, as shown in Figure 3-6. Pay attention to fastening through the satin balls with plastic ribbon to make it easy to paste them on cars.

图3-5 花泥的固定
Figure 3-5 fixation of flower mud

图3-6 花车
Figure 3-6 Float

（2）车门花

车门花最简单的制作方法是把单面花束（不包装）固定在车门上。车门处也可以粘贴大型的丝带花。固定车门花时，可以将花用透明胶条粘贴在车门把手周围，注意不要遮挡反光镜，以免影响行车安全。

（3）镶边花

鲜花镶边也称为花车车链，可以使用丝带镶边或直接用鲜花镶边，也可以两者结合使用，如图3-7所示。鲜花花链的制作很简单，就是把鲜花用胶带粘贴在汽车两边，注意镶边时应打开车门，以免胶带把车门粘住。也可以选择藤蔓植物，沿着车体边缘一点点地缠绕在小吸盘上。

（4）注意事项

①吸盘质量：在购买花车吸盘时一定要注意质量，仔细挑选，否则吸盘掉下来，花车的制作就失败了。

②色彩问题：对于黑色汽车，国内习惯用红色月季，如再配上些粉色和黄色的花会更亮丽一些。

(2) Door flower

The simplest method of making door flower is to fix the single side bouquet (unpackaged) on the door. Large ribbon flowers can also be pasted on the door. Fix the flower on the door handle with scotch tape, be careful not to cover the reflective mirror, lest affect driving safety.

(3) Embroidered flower

Embroidered flower is also called float flower chain which can be made of ribbons or fresh flowers or a combination of them, as shown in Figure 3-7. It is very simple to make fresh flower chain. Paste fresh flowers on both sides of the car with glue. Open the door of the car when embroidering to avoid the glue sticking on the car door.

(4) Cautions

① Quality of sucker: When buying float suckers, one must focus on the quality and select them carefully. Poor-quality sucks are easy to fall down, and the decoration of float would fail.

② Color problem: Red *Rosa rugosa* are usually used for the black car in China, which will be much brighter if pink and yellow flowers are also used together.

图3-7 花车车链
Figure 3-7 Float Flower chain

③时间问题：中国人接亲的时间依各地风俗不同，都有特别约定，所以花店应把做花车的时间控制在1h以内，车头花及镶边用的鲜花，应在用车的前一天晚上做好，第二天做花车时粘固即可。

④行车安全：从行车安全考虑，车头花的高度应低于30cm，可向横宽发展，在反光镜处不要粘花。很多花车车队为防止在路上走散，常常在车两侧挂上气球，将拴气球的绳子掩于车窗上即可固定，注意气球不要遮挡反光镜和司机视线。另外，做花车时需用剪子、刀子，同时花泥吸盘上还有铁把，注意不要划伤车身。

3.1.3 胸花

在婚礼中，新人要佩带胸花。胸花选用的花材要与新娘的头花及手捧花所用的花材一致。胸花常用的花材有月季、香石竹、石斛兰等。制作胸花时，首先把花头穿上绿铁丝，以防止断头，可以将绿铁丝的尖端弯一个钩，然后穿入花头，花头处理好后，把衬叶（如石松、蓬莱松、黄杨等）修剪成扇形作为后衬。叶子中间应高一些、稍密一些，接着在叶前衬上一些满天星或情人草等填充花材。然后，将穿好铁丝的花朵与叶子、情人草组合在一起，缠上绿色胶条，如图3-8、图3-9所示。最后，将与胸花宽度相同的丝带花结系于花头下边。注意女士别胸花的正确位置为肩以下、胸以上，男士为西服上衣口袋的位置。

③ Time problem: The time for picking up bride is different depending on the local customs in China with special agreement, so the florist's shop should time the float making within one hour. Front flower and embroidered flower should be completed before the night when using the car. Stick them on the car the next day.

④ Traffic safety: For the sake of safety, the height of front flower should be less than 30 cm. The front flowers can be arranged horizontally and don't stick flowers in the reflective mirror. Many floats team often hang up balloons on both sides of the car lest straying on the road. The rope tied to balloons can be fixed on the window, avoiding covering the reflective mirror and the driver's sight. In addition, scissors, knives, flower mud suckers and iron handles are necessary when making floats, and pay attention not to scratch the car.

3.1.3 Brooch

In a wedding, the couple will wear brooches. The floral materials selected to make the brooches must match those of the hand bouquet. Materials commonly used for brooches involve *Rosa chinensis*s, *Dianthus caryophyllus*, *Dendrobium nobile,* and etc. When making a brooch, you should pierce a green wire into the flower head to prevent breaking. The green wire can be bent into a hook and then pierce through the flower head. After that, cut and scallop the leaves (e.g. *Lycopodium japonicum* (clubmosses), *Asparagus mgriocladus* (asparagus myrioeladus), *Buxus sinica* (boxwood), and etc.) as back foil. The middle part of the leaves should be a bit taller and thicker. The front can be foiled with filling flower materials such as *Gypsophila paniculata* or *Codariocalyx motorius*. Then, combine the flowers pierced by wire, leaves and *Codariocalyx motorius* together and wind them with green glues tape, as shown in Figure 3-8 and Figure 3-9. At the end, ribbon knots with the same width as the brooch will be tied under the flower heads. Mind that the correct position for lady's brooch is below the shoulder and above the chest, whereas that for men is the position of the jacket pocket.

图3-8　胸花设计　　图3-9　胸花的固定
Figure 3-8　Brooch design　　Figure 3-9　Fixation of brooch

3.1.4　撒花

一般用玫瑰的花瓣作为撒花材料。当新人走入婚礼庆典大厅主道上时，撒花人员可以分2~3次，把手中的鲜花经空中抛向新人，不要提前撒，也不要追着撒，从新人入场走过鲜花门开始，在所走过的红地毯上，不断有人抛撒鲜花，具有时间的延续性。

3.2　庆典用花

3.2.1　制作特点

不同国家、不同民族对于喜庆场合的花色、风格等都有不同的传统习惯。就我国而言，喜庆场合常以红色、黄色、粉色等暖色系作为主要装饰色彩，寓意生意兴隆、财源广进、前程似锦等美好祝愿。如

3.1.4　Strewing Flower

Rosa rugosa petals are usually used as strewing flower material. When the new couple walks into main aisle of the ceremony hall, people responsible for strewing flowers will throw rose petals to the couple for 2 to 3 times. Don't strew early and chase after the couple. When the new couple passes through the flower door and walks through the red carpet, some people scatter flowers constantly onto the red carpet with the continuity of time.

3.2　Celebration Flower

3.2.1　Features

There are different traditions for color and style in celebrations in different countries and nationalities. In our country, China, warm colors such as red, yellow, pink, and etc. are the main decorative colors in celebrations which symbolize best wishes as the thriving business, plentiful treasures and bright future, and etc. For example, in the opening ceremony,

开业庆典，港台地区庆祝开业更多的是在一些铁艺架子上插花。而内地的开业庆典一般送开业花篮，规格有 1m、1.8m、2m 等，还可以在花篮上插制腰花。这种开业花篮花朵繁茂，色彩艳丽，如果摆放在门口，显得热闹喜庆，如图 3-10 所示。开业花篮也有插成四面观的，如图 3-11 所示。

花篮上应缀条幅，条幅一般为红底黄字。开业花篮条幅的一般写法为"某某公司开业大吉"，另一联为"某某公司敬贺"。

people in Hong Kong and Taiwan arrange flowers on the iron art shelf, whereas people in the mainland generally send opening basket of flowers in the opening ceremony, with specification of 1 m, 1.8 m, 2 m height, and etc. Flowers also can be inserted in middle of the basket. The opening flower baskets feature blooming flowers and gorgeous color. If they are put in the doorway, they can enhance lively and festival atmosphere, as shown in Figure 3-10. Opening flower baskets are also inserted into a view of all sides, as shown in Figure 3-11.

Flower baskets should be decorated with banners, which are commonly featured with yellow words and red bottom. Opening basket banners generally read "Great luck to ××× company's opening", another one "Congratulations from ××× company".

图3-10　庆典花篮
Figure 3-10　Celebration flower baskets

图3-11　四面观庆典花篮
Figure 3-11　All-side celebration flower basket

3.2.2 庆典花篮

在插制开业花篮时常用散尾葵或鱼尾葵叶片打底（图3-10），再插入菊花、牡丹、万年青、向日葵、月季、百合、花烛、鹤望兰、唐菖蒲、香石竹、满天星等。其制作方法如下：首先，在花篮中放入塑料纸防止花泥漏水，再在花篮中放入花泥，用绳子绑好。为防止绳子将花泥勒碎，绑绳的地方要垫上小棍。如果花篮要经过远距离运输，最好在花泥上面再罩上铁丝网固定。在花篮的中间部位，用绳子绑好半块花泥后紧固。在绑好花泥的花篮中间偏后的位置插入鱼尾葵叶，叶子前面可以插入线条花材，定出高度后插黄菊花。插入时注意等距，线条圆顺。黄菊花前面再插一排红色香石竹和切花月季，以加大颜色反差。再一排排依次插红色及黄色的鲜花，每层花应比上一层矮一个花头，直至插到花篮中间最低处为止。上面的花插好后，在已绑好的腰花花泥上先插入一圈鱼尾葵叶，叶子上面按等距插入黄金鸟，再插入一圈菊花、一圈红色香石竹和另一圈菊花，一圈圈插制鲜花，直至插到腰花的中心点。最后，在花篮中填配叶，完全遮盖住花泥。如果在花篮中插制百合、花烛等花，应插于花篮中间的焦点位置。

3.2.2 Celebration Flower Basket

When making flower baskets for opening ceremonies (Figure 3-10), we often use leaves of *Chrysalidocarpus lutescens*(chrysalidocarpus lutescens) or *Caryota ochlandra* (Caryota ochlandra) as the base, and then insert *Dendranthema morifolium, Paeonia suffruticosa, Rohdea japonica, Helianthus annuus, Rosa chinensis, Lilium* spp., *Anthurium* spp., *Strelitzia reginae, Dianthus caryophyllus, Gladiolus gandavensis, G.Paniculata*, and etc. The making method is as follows: first, put plastic in basket to prevent flower mud from leaking, then put flower mud in basket and tie up with rope. To prevent the rope from tightening flower mud, stack up small stick to the lashing place. If flower basket is transported during a long distance, it is advisable to hood wire entanglement on the flower mud. In the middle of the flower basket, tie half flower mud with a rope, Insert *Caryota ochlandr* in center-back of the flower basket tied with flower mud. Line material can be inserted in front of the leaves; insert yellow *Dendranthema morifolium* after the height is set. It should be equidistant and linear when inserting. Insert a row of red *Dianthus caryophyllus* and cut *Rosa chinensis* in front of yellow *Dendranthema morifolium* in order to increase color contrast. Then insert red and yellow flowers in turn, each layer of flowers should be shorter than the former, until they get into the lowest place of the middle of flower basket. After the flowers are arranged in the upper place, insert a round of *Caryota ochlandr* in the flower mud in the center of the flower basket. Insert heliconia on the leaves with equidistance, and then insert a circle of *Dendranthema morifolium*, a circle of red *Dianthus caryophyllus* and a *Dendranthema morifolium*. Flowers are inserted circle by circle till they get to the center of the flower basket. Finally, fill matching leaves in flower basket to cover flower mud completely. If *Lilium* spp., and *Anthurium* spp. are inserted in the flower basket, they should be put in the focus place in the middle of the flower basket.

3.3 丧事用花

丧事用花一般为黄色与白色花组成的图案、花篮、花束或者花圈，如图3-12所示。由于黄色和白色配在一起太显苍白，还可加入紫色的花卉，如勿忘我、石斛兰等，让花篮更漂亮。此外，马蹄莲、月季、孔雀草、郁金香、香石竹、百合、满天星、文竹、松柏枝等都可以用于丧事插花，但要注意花朵的颜色应素雅。丧事用花中，白菊花的应用最为广泛。

高龄老人去世，在国内称为喜丧，如经家人同意，喜丧用花可加入少许粉色和红色的鲜花。这种花篮以白菊花为主，同时插入了很多粉掌，适合于喜丧的场合使用。用鲜花制作的花圈在丧事中比较常用，下面介绍一下它的插制方法。

3.3 Flower Arrangement for Funeral Affairs

Flower arrangement for funeral affairs are generally flower baskets, bouquets and wreaths made of yellow and white flowers, as shown in Figure 3-12, of which yellow and white chrysanthemum are preferred. Because the match of yellow and white seems too pale, we can add purple flowers such as *Myosotis silvatica*, *Dendrobium nobile* and etc. to make flower baskets more beautiful. In addition, *Zantedeschia aethiopica*, *Rosa chinensis*, Tagetes patula, *Tulipa gesneriana*, *Dianthus caryophyllus*, *Lilium* spp., *Gypsophila paniculata*, *Asparagus setaceus*, *Pinus* spp. branches and etc. can be used for the funeral flower arrangement, but the color of the flowers should be simple but elegant. White *Dendranthema morifolium*. are the most widely used flowers for funeral affairs.

The occasion when a very elderly person died is called a happy funeral in China. With the approval of the family, slightly pink and red flowers can be used in the funeral. The basket is made mainly of white Dendranthema morifolium with pink *Anthurium* spp. inserted at the same time, which is suitable for happy funeral. Wreaths made of fresh flowers are frequently used in the funeral; the making methods are as follows.

图3-12 花圈
Figure 3-12 Wreaths

3.3.1 花圈

首先，根据作品需要选择不同的花圈架子，把整块花泥横着切开，将每块花泥紧挨着摆满花圈，用绳绑紧。绑绳处垫上花棍，以免绳子将花泥勒碎。固定花泥也可使用丝网。固定好花泥后，将要插入的花枝剪短，花头朝上一枝一枝紧挨着插满花圈，花圈插好后，用3根竹竿绑成支架。竹竿长度就是花圈的高度，竹竿的尖端可以用刀削尖，以便悬挂花圈。在绑扎竹竿前，可先在竹竿上面裹上白色或黄色丝带。然后，将3根竹竿立起，在其交叉点用绳子绑好固定，使它们能够立住。最后，以竹竿交叉点的绑绳处为界，在每根竹竿交叉点的上面和下面分别绕紧绳套，以防止挂上花圈后支架打滑。最常见的花圈是一圈一圈插制的，即插好一圈白色花后，再插入一圈黄色花，再配上绿草或点缀一些紫色花，中间预留空间嵌上"奠"字。花圈也可以插制出许多不同的样式，花圈所用条幅为白底黑字，用黑墨书写，一联为"沉痛悼念某某"，另一联为"某某敬"。

3.3.2 花篮花束

丧礼花篮常见的造型有单层式、双层式、多层式、灵前小花篮等。如单层式丧礼花篮可选用三角架来做支撑，因三角架有一定的高度，再蒙上纱绸布，用花量少，可用较高档的花，此种花篮便于运输、较轻巧。

3.3.1 Wreath

First choose different wreath frames according to what the work needs, slice the whole piece of flower mud sidewards, then lay each mud around the wreath and tighten it with rope. The place tied with rope should be stacked up with flower branches lest the rope smashes mud. Silk fabric can be used to fix mud. After flower mud is fixed, the flower branches inserted should be cut short. Branches of flowers are stick around the wreath one next to another whose heads are upwards. After finishing the wreath, we use 3 bamboo poles to bind the holder. The length of bamboo is that of wreath. The tip of bamboo can be sharpened with a knife so as to hang wreath. Before binding bamboo, we can wrap white or yellow ribbon around the bamboo. Then, erect 3 bamboo ropes and tie up its crossing with rope so as to enable them to stand. Then take the crossing place bound with rope as the boundary, roll bails tightly at the top and bottom of each bamboo pole intersection for fear that the holder slide after hanging wreaths. The most common wreaths are arranged circle by circle, namely a round of white flowers are inserted and another round of yellow flowers are added. Some green grass are decorated on the wreath or it is interspersed with purple flowers. In the middle of the wreath a space is reserved for and embedded by the character "奠", which means making offerings to the spirits of the dead. Wreaths can also be made in many different styles, as is shown in the picture. Banners used for wreath are of white ground and black text, written in black ink, one banner is "Deep mourning to ×××", another one is "Offered by ×××".

3.3.2 Bouquets in Basket

The common shapes of flower baskets for funeral include single-deck, double-deck, multilayer and small flower basket in front of coffin. Single-deck flower baskets can be supported by tripod, because tripod has a certain height, and is then covered with yarn silk and inserted with a few flowers of higher-grade. This type of flower basket is convenient for transportation and is light and handy.

丧礼花束一般有两种用途，一是追思会和遗体告别时，敬献给亡者，表示对亡者的敬仰和深切的缅怀；二是上坟扫墓时带上一束，敬献坟前，以示哀悼，花束大多以单面观造型为主。制作上除花材的使用特殊外，方法与普通的花篮、花束制作大多相同。

3.3.3 摆花

丧礼摆花的形式主要有礼厅布置、遗像衬花、遗体花坛、遗像铺花、家庭灵堂。如礼厅布置，礼厅是丧礼中最为主要的场所，一般是进行遗体告别仪式或者追悼会的地方，是家属见逝者最后一面的地点，是设计的重点。设计分为礼厅门口和礼厅内两部分。礼厅大门一般会悬挂黑色横幅，两侧挂挽联。插花应对称设计，颜色以白绿为主色系，一般用白菊花，配材尽量简单，突出稳重感，还可以在作品之间用白纱幔或白布相连，或挂悼念横幅，以更好地营造丧礼气氛。礼厅内部摆花除配合整体布景外，应考虑功能需求，要留出遗体告别仪式的过道，可以用花篮、花圈等进行引导。

3.4 会议和宴会用花

3.4.1 制作特点

会议室插花布置的形式以低矮、匍匐形、宜四面观赏的西方式插花为主，在沙发转角处或靠墙处茶几上也可用东方式插花。

Funeral bouquets generally have two kinds of usage. One is offering to the deceased when we say goodbye and pay tribute to the dead, showing respect and deep memory to the deceased. Another is to bring a bunch of bouquet, lay it in front of the grave for mourning. The main shape of the bouquet is one-side view. Except for special material used for making such bouquets, methods are mostly the same as those for making common flower baskets.

3.3.3 Flower Arrangement

The form of funeral flower arrangement is mainly hall layout, accessory flowers for dead, body parterre, laying flowers or deadee, the family mourning hall. Take hall layout as an example. Hall is the most primary place in the funeral, which is usually to say goodbye or hold memorial service. It is the place where families see the dead for the last time and it is the key in the design. It is divided into the gateway and the inner of the ceremony hall when designing. A black banner will generally be hung upper the door, elegiac couplets are hung on both sides of the door. Flower arrangement should be designed with symmetry. Color is given priority to white and green system, generally with white chrysanthemum. Materials should be as simple as possible, highlighting sedate sense. These works can also be connected with white gauze curtain or white cloth, or hanging a memorial banner to create a better atmosphere of the funeral. In addition to coordinating the whole setting in ceremony hall, we should consider the functional requirements and set aside aisle for mourners to say goodbye to the dead, which can be guided with flowers, wreaths and others.

3.4 Flowers for Conferences and Banquets

3.4.1 Features

Flower arrangement for meeting room mainly features low, creeping and western form suitable for viewing all sides, as is shown in the picture. In the corner of sofa or the tea table on a wall, eastern flower arrangement can also be used.

无论哪种插花形式，一是花要新鲜、艳丽、盛开，二是花无异味或浓香，三是花的高度切忌遮挡与会者发言或交谈的视线。插花的规格依会议的级别而定，一般会议只是在主席台或中间（圆桌会议）插制一至数盆不等，而高级会议在一般会议布置的基础上，不但花要高档，而且数量也要增多。如签到处、贵宾休息处、会议室四角等处都应布置。

3.4.2 常用形式

常用的形式有会议桌花、嘉宾胸花、签到台花艺、演讲台花、迎宾牌花艺等。如会议桌花是放于会议及宴会桌上的插花，一般为圆形或椭圆形，高度不应超过30cm，以免遮挡视线，宽度以不妨碍进餐或开会为宜，如图3-13所示。

插花的规格依会议的级别而定。花泥要处理干净，保持整洁，花器要做好防水处理，以免漏水沾染木质桌面。无主席台的圆形或椭圆形会议桌上以放置圆形、椭圆形或长方形桌花为宜，如果没有主席台，那么主席台第一排会议桌中间位置桌花以平铺形或下垂形为宜，如图3-14所示。

No matter what flower arrangement form, flowers must be fresh, gorgeous and blooming. Besides, flowers must be free from extraneous odour or aroma. The third one is that the height of the flower can't block conventioners' sight when speaking or talking. The specification of the flower arrangement should be in accordance with the level of meeting. Arrange a pot of flower or several pots in the platform or middle (round-table conference) for ordinary meeting, whereas arrange high-grade flowers and increase the number of them for high-level meeting on the basis of the general meeting arrangement. For example, flowers should be arranged in places of registration, the public room for distinguished guests and the four corners of conference room.

3.4.2 Common Styles

Common styles include board flower, guest brooch, registration desk floriculture, podium flower, guest greeting board floriculture and etc. For example, board flower is put on the table of meeting and banquet, it is usually round or oval, its height should not exceed 30 cm, so as not to block the sight and its width should not interfere with dinner or meeting, as shown in Figure 3-13.

The specification of the flower arrangement should be in accordance with the level of meeting. Flower mud must be clean and tidy. Floral vase must be waterproof so as not to leak and stain woodiness desktop. It is advisable to put round, oval or rectangular table flowers on the round or oval table without rostrum. If there is no rostrum, it is advisable to put flat or drooping flower on the middle table in the first row, as shown in Figure 3-14.

图3-13 会议桌花
Figure 3-13 Board flower

图3-14 下垂形会议桌花
Figure 3-14 Drooping board flower

3.5 生日用花

3.5.1 制作特点

不同的人过生日，应送不同的鲜花。老年人过生日可送以松枝、鹤望兰、菊花（有长寿之意）等花材插制的花篮，取松鹤延年之意。给老人祝寿的鲜花还可以搭配万年青、竹子、石斛兰、金鱼草等。需要注意的是，菊花虽有长寿之意，但很多人认为它是丧事用花，所以，在使用菊花时最好征求一下顾客的意见。

爱人过生日可以送切花月季、百合、郁金香、马蹄莲、勿忘我等。如果是丈夫过生日，还可以送以非洲菊为主的花篮或花束，因为非洲菊又名扶郎花，有扶助郎君之意，送给丈夫最为合适，如图3-15所示。

母亲生日送以香石竹为主的花篮、花束，如图3-16所示。

3.5 Birthday Flower Arrangement

3.5.1 Features

Different flowers should go to different people's birthday party. For the elderly's birthday, we can send flower baskets made of *Pinus* spp., *Strelitzia reginae*, *Dendranthema morifolium* (with the implied meaning of longevity), which implies having longevity as pines and cranes. Flowers for the elderly's birthday can also be matched with the *Rohdea japonica*, Bambusoideae, *Dendrobium nobile*, *Antirrhinum majus* and etc. It is important to note that although *Dendranthema morifolium*. implies longevity, a lot of people think it is the flower for the funeral, so, when using *Dendranthema morifolium*.it's better to ask for advice from the customers.

For lovers' birthday, we can send cut *Rosa chinensis*, *Lilium* spp., *Tulipa gesneriana*, *Zantedeschia aethiopica*, *Myosotis silvatica* and etc. If it is for husband's birthday, we can also send flower baskets or bouquets mainly made of *Gerbera jamesonii*, because *Gerbera jamesonii* is also known as "Fulang" flower in Chinese characters with the implied meaning of helping husband. It is the most appropriate flower to send husband, as shown in Figure 3-15.

For mother's birthday, it is advisable to send flower baskets or bouquets mainly made of *Dianthus caryophyllus*, as shown in Figure 3-16.

图3-15 以非洲菊为主要花材的插花
Figure 3-15 Flower arrangement with gerbera as main floral material

图3-16 香石竹为主的花篮
Figure 3-16 Flower basket with carnation as main floral material

3.5.2 常用形式

生日用花常用的形式有花束、花篮、个性形式等。如花束的制作，一般由花体、手柄和装饰三部分组成。常用形状有单面观赏花束、扇形花束、直线花束、四面观赏花束、半球花束、放射形花束、球形花束、不对称组群花束等。单面观赏花束的种类又很多，有很大的可变性，如扇形、直线形等。如图3-17所示为常见四面观花束，图3-18所示为单面观花束。

3.5.2 Common Styles

Styles commonly used include bouquet, flower basket and personality, etc. For example, birthday bouquet generally consists of three parts, namely, squiggles, handle and decoration. Common shapes include single-side ornamental bouquets, fan-shaped bouquets, linear bouquets, all-side ornamental bouquets, hemisphere bouquets, radiate bouquets, balloons bouquets, asymmetric group bouquets. Single-side ornamental bouquets have many varieties and great variability, such as fan and linear. Different materials should be chosen based on different objects for birthday flowers. Figure 3-17 is an all-side ornamental bouquets, and Figure 3-18 is a single-side ornamental bouquet.

图3-17　花束
Figure 3-17　Bouquet

图3-18　生日花束
Figure 3-18　Birthday bouquet

3.6 探望病人用花

3.6.1 制作特点

探望病人用花在品种和颜色方面有讲究，花束适宜选择相对鲜艳柔和的色彩，如唐菖蒲、香石竹、花烛、兰花、水仙、马蹄莲等。也可配以文竹、满天星或石松，以祝愿贵体早日康复。或选用病人平时喜欢的种类，有利于病人怡情养性，早日康复。

给病人送花也有一些禁忌，如探望病人时不要送整盆的花，以免病人误会为久病成根。香味很浓的花对术后的病人不利，易引起咳嗽；颜色太浓艳的花，会刺激病人的神经，激发烦躁情绪；山茶容易落蕾，被认为不吉利。还要注意忌送的数目：4，9，13。

3.6.2 常用形式

常用的形式有花束、花篮，如探望病人的花篮有元宝状花篮、荷叶边花篮、筒状花篮、浅口花篮等，插花造型有四面观赏型和单面观赏型。花篮用花除注意医院的特殊性外，还要顾及病人的喜好，制作手法和普通花篮的要求一致。

3.7 其他场合用花

除以上应用外，还有居家装饰用花、宾馆装饰用花、商业装饰用花、公务装饰用花等。如家居装饰用花的基本形式丰富多样，一般用花篮、花束、艺术插花等个性形式，用花上也不固定，根据家庭和个人的喜好以及花材情况灵活应用，只要花材新鲜，

3.6 Flowers for Visiting Patients

3.6.1 Features

If you are going to the hospital to visit a patient, you should be particular about the varieties and color of flowers. It is appropriate to choose bright *Gladiolus gandavensis*, *Dianthus caryophyllus*, *Anthurium* spp., *Cymbidium* ssp., *Narcissus tazetta*, *Zantedeschia aethiopica* and etc. In addition, matched with *Asparagus setaceus*, *Gypsophila paniculata*or, *Lycopodium japonicum* is a good choice for a patient in order to wish him/her to recover soon. Or choose patients' favorite variety so as to benefit the patient's mind and character, and help the patient recover soon.

There are a lot of taboos for sending flowers to a patient. Do not send a whole pot of flowers lest patient misunderstands that it symbolizes for long illness into the root. Flowers with thick fragrance are against surgical patients, which are inclined to cause cough; Flower with thick color can stimulate the patient's nerve and arouse irritable mood; *Camellia japonica*, whose bud falls easily, is considered bad luck. The numbers such as 4, 9, 13 are forbidden.

3.6.2 Common Styles

Common styles include bouquets and flower baskets. For example, shapes of baskets for visiting patients include shoe-shaped ingot, falbala, tubular, shallow mouth and etc. Shapes of flower arrangement include those being all-round ornamental and single-side ornamental. The particularity of the hospital and the patient's preferences should be taken into account when we choose flowers for baskets. Making technique is in accordance with the requirements of the common flower baskets.

3.7 Flower Arrangement on Other Occasions

In addition to the above mentioned occasions for flower arrangement, other occasions include household decorative

图3-19 彩菊花篮
Figure 3-19 Basket of colored chrysanthemums

flowers, hotel decorative flowers, commercial decorative flowers, official business decorative flowers and so on. For example, the basic form of household decorative flowers is rich and varied. The common form is of personality with flower baskets, bouquet and art flower arrangement. There is no fixation of using flowers, depending on family and personal preferences and flower material to apply flexibly. As long as the flowers are fresh, the owner loves,

主人喜欢，能让人赏心悦目，美化环境就可以，如图3-19所示。另如在客厅落地窗前的陶瓷鱼缸里用重物固定花泥，插上荷花，配上荷叶，就是一幅非常生动的花鱼共赏插花作品。

which can please people and beautify the environment, it is alright, as shown in Figure 3-19. If we put flower mud in the fish porcelain jar fixed with heavy objects in the front of the floor window in the living room, plugged with lotus, covered with lotus leaf, it becomes a very vivid flower arrangement work with both flower and fish appreciation.

第4章 插花命题及花文化内涵

Chapter 4 Propositions of Artistic Flower Arrangement and the Connotations of Flower Culture

4.1 插花的命题立意

4.1.1 命题立意的特点

（1）意境美是插花的组成部分

意境，是艺术家审美的再现，是与生活形象融为一体而形成的艺术境界，在有限的作品中表达无限而深远的内涵。插花艺术的意境之美，主要靠花材的特性、独具匠心的造型、配件的恰当使用、花器的巧妙选择、色彩的合理搭配等手段而营造出的一种景外之景，意外之意。意境之美，是东方式插花艺术的最高境界，也是现代艺术插花追求的最高目标。

（2）重意境是中国插花艺术的一大特点

东方式插花崇尚自然、师法自然并高于自然，善于利用自然花材的美来娱人、感人。不仅注重花材的形态美和色彩美，而且更注重花材所表达的意境美。在插花中注重花材的人格化意义，注重花材的文化因素，赋予作品深刻的寓意，借插花来表达作者的精神境界。

（3）意境无限定

插花创作中的意境营造是根据作者想要表达的主题而定。使用的花材不同，创作的立意不同，意境就不一样。因此意境的设定无限定。

4.1.2 命题立意的要求

①切题；

4.1 Propositions of Artistic Flower Arrangement

4.1.1 Characteristics of Proposition

(1) Beauty of conception being a part of artistic flower arrangement

Artistic conception is the representation of an artist's aesthetics, and is also the artistic state resulting from the integration with the image of life. It expresses profound and unlimited connotations in limited works. The beauty of conception in artistic flower arrangement depends mainly on the scene and connotation created by the characteristics of floral materials, ingenuity of shaping, proper use of accessories, clever selection of vessel, and reasonable combination of colors. Beauty of conception is the tidemark of oriental flower arrangement, and the highest end of contemporary artistic flower arrangement.

(2) Focus on conception being a major feature of the art of Chinese flower arrangement

Oriental flower arrangement is nature-respecting, which learns from nature and exceeds nature. It is good at making use of natural floral materials to entertain and impress people. Oriental flower arrangement lays emphasis not only on the beauty of shape and color, but also on the beauty of conception conveyed by floral materials. In the arrangement, great attention is paid to the personification and cultural elements of floral materials to endow the arrangement with deep moral, and to express the arranger's spiritual condition.

(3) Artistic conception being unlimited

The artistic conception created in flower arrangement is based on the theme the arranger tries to convey. Different floral materials require different propositions, and consequently different conceptions. Therefore, the artistic conception in flower arrangement is unlimited.

4.1.2 Requirement for Proposition

① Being to the point;

② 积极健康；
③ 含蓄；
④ 有文化品味。

4.1.3 命题立意的基本方法

（1）意在笔先

先出题目，再如作文一样，用花材将主题表现出来。

（2）意随景出

事先没有命题，在插制过程中灵感所至，创造出别具特色的好作品。

4.1.4 命题立意的具体方法

（1）借助主要花材的寓意或形状命题

中国传统插花善于利用各种花材的形态质地特点和习性气质赋予其种种美好的象征意义，并加以人格化来表达主题、情感和意趣。如将松、竹、梅插于一瓶，取名"岁寒三友"，以表达朋友间患难与共的真挚友情和刚正不阿的人生态度；又如荷花出污泥而不染，香远益清，人们赋予其品格高尚、端庄圣洁的寓意。结合插花作品的主题，巧妙运用花材的寓意进行创作，常引人遐思联想。因此，了解和熟悉花木的各种特性，以及人们赋予花木的精神，对于恰当表达作品意境非常重要。

（2）借助造型命题

利用插花作品造型命题，依形写神，以假当真，运用形象思维去展开联想，比拟真景，以其神态恰当命题。

② Being positive and healthy;
③ Being implicit;
④ Being culturally tasteful.

4.1.3 Basic Methods of Proposition

(1) Theme prior to arranging

Like writing an essay, flower arrangers should set the theme first, and then express it through floral materials.

(2) Theme along with arranging

The inspiration may occur in the process of arrangement if being short of a theme beforehand, and a good flower arrangement work may be created.

4.1.4 Specific Methods of Proposition

(1) Proposing according to the moral and shape of floral materials

Chinese traditional flower arrangement makes good use of the shapes, textures, habits and temperaments of different floral materials to create fine symbolic meanings, and personify them to convey the theme, emotion and charm. For example, *pinus* spp., *Bambusoideae* and *Armeniaca* spp. blossom are arranged in one vase, and named "three cold-weather friends" to express genuine friendship and upright attitude towards life. For another example, *Nelumbo nucifera* rise from dirty mud unsoiledly with delicate fragrance, and are endowed with the moral of nobility and purity. Clever use of the moral of floral materials in combination with the theme of the flower arrangement can often arouse people's association and imagination. Thus, in order to express the artistic conception of the flower arrangement works, it is of great importance to learn about and get familiar with the characteristics of various flowers and trees, as well as the spirits people give them.

(2) Proposing according to the shape of the arrangement

The arranger can apply image thinking to launch association according to the shape and mien of the flower arrangement works, and make suitable propositions.

（3）借助色彩命题

色彩是插花艺术中非常重要的因素，对于意境的形成至关重要。在很多插花比赛中，选手们常用闪亮的银色、深沉而富于变化的蓝色、神秘的紫色来表现对世界的神往与期盼；用鲜丽的红、橙等暖色表现欢乐的心情。而当一幅作品出现粗重的线条和大片灰、褐、棕、黑等色调时，它要带给人们的肯定是深深的反省和沉重的思考。

（4）借助器具命题

器具包括容器和配件，它们是插花作品的重要组成部分，是艺术构思的一部分。器具的选择恰当与否，直接影响到作品的成败。

（5）借助综合条件命题

根据作品中花材、花器或配件等所具有的特殊含义，借古诗、名诗佳句、典故、景观等综合条件命题。

4.1.5 提高命题立意素养的方法

（1）提高文化水准

插花艺术是一门综合性艺术，其涉及植物学、花卉学、美学、文学、素描、绘画、色彩学等相关学科。因此，若要提高命题立意素养，必须不断提高自身的文化修养。如此才能丰富作品的内涵，表达美的意境。

（2）掌握花材知识

插花创作最基本的材料就是花材，掌握较全面的花材知识，有利于把握花材的特性，领会花材的风姿神韵，更好地表现花材。

(3) Proposing according to the color

Color is a very important element in the art of flower arrangement, and is crucial to the formation of artistic conception. In many flower arrangement contests, the contestants often use the shiny silver, deep but changeable blue, and mysterious purple to express the expectations and yearnings for the world, while using warm colors like bright red and orange to express a happy mood. If an arrangement uses heavy lines and big patches of grey, sepia, brown and / or black, what it wants to provoke must be deep introspection and meditation.

(4) Proposing according to utensil

Utensil is an indispensable part in flower arrangement, which includes containers and accessories. It is also a part of artistic conception. The choice of utensils will directly affect the success of flower arrangement works.

(5) Proposing according to comprehensive conditions

The arranger can use the comprehensive propositional conditions of ancient poems, well-turned phrases, literary quotations and landscapes to make propositions according to the special meanings of the floral materials, vessels and accessories used in the flower arrangement.

4.1.5 Ways to Improve the Accomplishments of Proposition

(1) Improving educational level

Flower arrangement is a comprehensive art, involving botany, floriculture, aesthetics, literature, sketch, painting, chromatology and other related disciplines. Therefore, in order to improve the accomplishments of propositions, flower arrangers need to keep improving their own educational accomplishments so as to enrich the connotation and express the beautiful conception of the works.

(2) Mastering the knowledge of floral materials

Floral materials are the most basic materials in flower arrangements. A good mastery of the comprehensive knowledge of floral materials is conductive to grasp the characteristics and appreciate the graceful charm of floral materials, and to manifest them better.

(3) Accumulating related knowledge

The art of flower arrangement in China has been developing fast in recent years, and many excellent talents and works are springing up. Flower arrangers should collect good works to make comparisons, to learn from other arranger's strong point and improve their own capability of creation and appreciation.

(4) Observing nature and society

Flower arrangement belongs to artistic creation, which requires a thorough observation of the society, and a deep involvement in life and nature to seek the inspiration and source of creation. Together with continuous thinking and practicing, the works can be at a higher artistic level both in form and in thoughts and feelings.

4.2 Appreciation of Artistic Flower Arrangement and Flower Culture

The blooming and fading of flowers are both aesthetics and philosophy. Flower is the reproductive organ of a plant, so in the natural world, it can usually affects people's subtle feelings about life most easily. Flowers not only possess the objective attribute of the beauty in substance, but also the subjective aesthetic from human. They are the objects and symbols of beauty, as well as the internalization of people's spirit beyond the reality. Therefore, enjoying flowers is always an important part of human culture.

Compared with potted flowers and garden plants, many artistic flower arrangements were made by cut flowers whose biggest feature is the separation from the mother plant. So, the biggest difference between the appreciation of cut flowers and other flowers is that cut flowers those are incomplete from the perspective of physiology, but from the aesthetic perspective, they express a tension of beauty in the incompleteness. For this reason, cut flower exhibits its particularity in application. First, it is time-limited. A flower cut from the mother plant is in a condition of injury with the nutrition and water supply cut off, which makes its aging much faster than potted flowers and

进行取舍，这一点是盆花和园林花木所不能及的。艺术插花在点缀人们生活的同时更带给人们自然的气息，它和人类的关系昭示着人与自然和谐共处的美妙。

4.2.1 花卉欣赏与花文化

文化是一个民族的灵魂和血脉，比如中国的花文化在一定程度上承载着中华民族的精神，体现着中华民族的认同感、归属感。我国各族人民皆有各具特色的赏花、用花习俗以寄托丰富的思想感情。

人们在养花、赏花的实践中，从花木身上得到启迪，又将大自然赋予花木的自然品性提炼上升为可以指导人进行社会活动的意识形态，直接作用于人的价值观、世界观和方法论。从花色、花韵中提炼的花情、花德，正是人类在推动自身社会文明进步中积累的宝贵精神财富。儒家的等级观念和道家的万物有生的个体思想，均交融于中国古人对花木的观赏和体验中，人们通过对花木的观赏寻求"天人的合一""人与自然的平衡"。因此，中国人历来强调人与自然的协调，强调个体对群体的适应，即"和"，且表现在人与花的辩证统一中。

garden plants. Second, it is more flexible in application. Packed cut flowers are convenient for both short and long distance transportation, and unlike potted flowers and garden plants, they can be partly adopted or discarded according to the needs in decoration arrangement. Artistic flower arrangement brings people the atmosphere of nature while decorating people's life. Its relationship with human beings declares the wonderfulness of the harmonious coexistence of human and nature.

4.2.1 Flower Appreciation and Flower Culture

Culture is the soul and bloodline of a nation. Chinese flower culture, as an example, carries the mental memory of Chinese nation, and embodies the sense of identity and belonging of Chinese nation. The ethnic groups in China have their own customs of flower appreciation and use, to which rich thoughts and feelings are attached.

In the practice of flower raising and enjoying, people can be enlightened by the flowers, then extract the quality of flowers endowed by nature and upgrade it to the ideology which can be used to guide people's social activities. This ideology can directly affect people's values, world outlook, and methodology. From the color and charm of flowers, people extract love and virtue, and this is exactly the precious spiritual wealth of human beings accumulated in the process of promoting the progress of social civilization. Confucian concept of hierarchy and Taoist individual ideology of hylozoism are both integrated in the appreciation and experience of flowers of ancient Chinese. They sought the "unity of human and nature" and the "balance between human and nature" via the appreciation of flowers. Therefore, Chinese people always put emphasis on "harmony", which means the coordination of man and nature, and the individual adaption to group. The "harmony" is also embodied in the dialectical unity of man and flower.

Now, the flower industry is no longer a mere planting industry. It is incorporating into the modern culture industry, which is a sign that the study of flower culture begins to promote the further interpenetration of horticultural science

花卉发展正由传统单纯的种植业向现代文化产业并轨，这标志着花文化的研究开始促进园艺科学与人文科学进一步相互渗透。人文学科的研究对象是人的精神世界和文化世界，它总是要设立一种理想人格的目标和典范，从而引导人们去思考人生的目的、意义、价值，追求人的完美。而花卉即兼具社会物质文明建设和精神文明建设的双重角色。人们通过花文化的传播和继承，促进花文化事业与产业的有机融合，最大限度地发挥花卉产业的经济实力；同时，传统花文化也承载着人类所宣扬的人文主义精神和民族文化的精髓，通过人与花的互动，在社会物质文明、精神文明以及生态文明的建设中给人以启迪，构建人与人和睦相处、人与自然和谐与共的地球文明。

4.2.2 花卉欣赏要点

花卉的美和世间万物的美一样都有其外在美和内在美，花卉的色、香、姿是外在美，而韵则是隐藏在花卉实体之中的内在美。赏花是以花之美，叩开心扉；以爱美之力，扫除虚假与邪恶，净化心灵，使精神世界开阔、乐观，少有烦恼与苦闷，从而保证身心健康，是以"修身养性"！赏花遵循的原则是：首先观其色，近而闻其香，细心赏其姿，终而品其韵。

（1）首先观其色

花卉是大自然的精灵，是美的化身。色彩是最容易给人感官刺激的要素之一。切花的色彩异常丰富，浓妆淡抹总相宜，人们秉性不同，喜爱也各异。

and humanity science. The object of the humanities is the spiritual world and cultural world of human beings, and it always tries to set an aim and example of ideal personality to guide people to think about the goal, the meaning, and the value of life, and seek perfection of mankind, while flower plays a dual role in the constructions of social material civilization and spiritual civilization. The propagation and inheritance of flower culture promote the integration of the cultural undertaking of flower and the flower industry, through which the full economic strength of the flower industry can be presented. Moreover, the traditional flower industry is also the carrier of the essence of humanism and national culture advocated by people. It gives people enlightenment in the constructions of social material civilization, spiritual civilization and ecological civilization through the interaction with people, and guide people to create a global civilization with harmony in human society, and between human and nature.

4.2.2 Gist of Flower Appreciation

Like all the other things in the world, flowers have their external beauty and inner beauty. The color, fragrance and appearance are the external beauty, while the charm is the inner beauty hidden in flowers. Flower appreciation is to open our hearts with the beauty of flowers, wipe out falsehood and evil, and purify our souls with the love of beauty, making our spiritual world open and optimistic with less annoyance and gloom to ensure our soundness in body and mind. In the appreciation of flowers, the principles we should follow are: color first, fragrance second, appearance third, and charm last.

(1) Color first

Flowers are spirits of nature, and the embodiment of beauty. Color is one of the elements easy to bring people sensory stimuli. Cut flowers are exceedingly rich in colors, and always charming with either light or deep colors. People with different dispositions may have different preferences.

（2）近而闻其香

花卉的香味以浓、清、远、久来评判，有的人喜欢浓郁的芬芳，有的人喜欢淡雅而弥久的清香。中国人赏花往往香重于色，神重于形，在香味方面认为茉莉当数人间第一香；兰则号为"香祖"，尊为国香。

（3）细心赏其姿

花的姿态有柔美和刚美之分，柔抒韵、刚抒气；既有婉和阴柔之情，又有稳重阳刚之气；静中有动，绵里藏刚。

（4）终而品其韵

花是美的化身，美是以真与善为条件而存在的，花草树木在文人雅士的笔下流淌出各具特色的精神风貌，也是人类赋予花卉真、善、美的精神内涵。花韵就是花的风度、品德和特性，是内在的美，真正的美，抽象的意境美。赏花求雅、赏花论韵，人们赏花若不谙花韵，则难入高雅境界。

爱美是人类的天赋，如果爱美而不追求真与善，就只能触及美的外表，达不到爱美赏花之目的。

4.2.3 花卉文化内涵在作品中的表现

一个传情的插花作品往往饱含作者丰富的思想感情，花材与花艺手法的搭配是表达这种感情的物质基础，而更为关键的则是蕴含在作品中的文化内涵。通过花语、花德来再现作者内心，让插花作品的创作达到一定的精神境界，配以恰当的名字或者创意介绍，更能引领大众对插花作品的深度解读和赏析。请从以下案例仔细体会花卉文化对作品寓意的提升。

(2) Fragrance second

The fragrance of flowers can be judged as heavy, light, far-reaching and lasting. Some people like a heavy fragrance, while some prefer a light and lasting one. In Chinese people's appreciation of flowers, fragrance is usually more important than color, and spirit more important than form. As to fragrance, they believe that *Jasminum sambac* has the most pleasant fragrance in the world, and *Cymbidium* spp., with the title of "National Fragrance", is honored as the ancestor of all fragrances.

(3) Appearance third

The appearance of flowers can be divided into soft beauty and hard beauty. Soft beauty gives expression to charm, while hard beauty gives expression to spirit. So, in the appreciation of flowers, you cannot find only the gentle and soft charm, but also the solid and steadfast spirit; motion in quiescence, and a firm character behind a gentle appearance.

(4) Charm last

Flower is an embodiment of beauty which exists on the basis of truth and goodness. Plants and flowers have their distinctive spirituality under the pens of refined scholars, which is also the spiritual connotation people want to endow flowers with. The charm of flowers refers to the grace, virtue and characteristics of flowers, which is the inner and true beauty, and the abstract beauty of conception. Appreciation of flowers is in pursuit of elegance, and pays attention to the charm of flowers. If one appreciates a flower without understanding its charm, he can hardly reach the realm of elegance.

The love of beauty is the nature of human beings. If we love beauty without the pursuit of truth and goodness, the love is superficial, and the aim of flower appreciation cannot be achieved.

4.2.3 Manifestation of Cultural Connotations of Flowers in Flower Arrangement Works

An emotion-expressing flower arrangement is usually full of the arranger's thoughts and feelings. The collocation

案例一：雷雨（图4-1）

　　作者：刘飞鸣、邬帆

　　主要花材：钢草、公主花、针垫花、苏铁雄花、地涌金莲、百合、马蹄莲、大花蕙兰。

　　作者创意：漫天翻滚的乌云，压得人透不过气来，劈空而下的闪电，暴雨倾盆而下，滋润着被烈日灼烤得发烫的大地，花草重又焕发生机，绽露出生命的色彩。

　　点评：作者以花草为素材，描绘出一幅形神具备的天象地貌，尤以钢草表现雷雨、龙柳表现闪电为神来之笔，产生了"迅雷不及掩耳，疾霆不暇掩目"的艺术感染力。该作品紧扣"人与自然"的主题，突出自然现象与生态平衡的关系，通过插花艺术的表达，警示人类应该有"尊重自然，保护生态"的观念。

of floral materials and arranging techniques is the material base to express such feelings, while the cultural connotation contained in the work is more crucial. It can better lead the audience to further interpret and appreciate the flower arrangement if the arranger's inner world is shown through the flower language and virtue to lift the work to a certain spiritual state, and a proper name or a creative introduction is given. Please appreciate the following works.

Case One: Thunderstorm (Figure 4-1)

　　Arrangers: Liu Feiming & Wu Fan, Nantong

　　Main floral materials: *Xanthorrhoea* spp. (steel grass), *Melastoma candidum* (tibouchina, princess flower), *Leucospermum cordifolium* (leucospermum), male flower of *Cycas revoluta* (cycas revolute), *Musella lasiocarpa* (musella lasiocarpa), *Lilium* spp., *Zantedeschia aethiopica* and *Cymbidium faberi* (cymbidium).

　　Originality: The dark clouds rolling in the sky smother people, and with a big flash of lightning, a heavy rainstorm pours down, moisturizing the hot earth scorched by the sun. Flowers and plants regain their vitality, and begin to show the color of life.

　　Comments: The arranger uses flowers and grasses as materials, and creates a picture with the unity of form and spirit. An ingenious stroke is that he uses *Xanthorrhoea* spp. to stand for thunderstorm, and Salix matsudana 'Tortuosa' (salix liouana) for lightning, which produces the artistic appeal of an oncoming storm. The work closely follows the subject of "human and nature", and highlights the relation between natural phenomena and ecological balance, warning people to hold the cultural belief of "Respecting Nature and Protecting Ecology" through the artistic expression of flower arrangement.

图4-1　《雷雨》（杨发顺摄）
Figure 4-1　*Thunderstorm* (Photographed by Yang Fashun)

案例二：百折不挠（图4-2）

　　作者：章红

　　主要花材：水葱、棕榈、百合。

　　作者创意：个人乃至社会要前进、要发展，必须具有不畏艰险、百折不挠的精神。通过植物材料的折线造型，不仅在形式上直点主题，而且重在表现一种精神和毅力。

　　点评：作品材料简洁、构图明快，百合（隐射"百"）从曲折的空间（隐射"折"）中挣扎向上，一朵含苞待放的花蕾直立在顶端，寓意着只有经历曲折，不畏艰险，才能脱颖而出。从正面点明"百折不挠"的精神品质。

Case Two: Perseverance (Figure 4-2)

Arranger: Zhang Hong

Main floral materials: *Scirpus tabernaemontani* (scirpus tabernaemontani), *Trachycarpus fortunei* (palm), *Lilium* spp..

Originality: To progress and develop, a person or even a society must have the spirit of perseverance and fearlessness. Through the fold line shaping of the plant materials, the work goes straight to the theme in form, and also puts emphasis on the expression of a kind of spirit and willpower.

Comments: With succinct materials and sprightly composition, the work throws light on the spirit of "perseverance" with a *Lilium* spp. in bud struggling upward in the zigzag space. It implies that only by experiencing twists and turns, and braving hardship and crisis, can one stand out from the crowd.

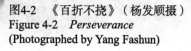

图4-2 《百折不挠》（杨发顺摄）
Figure 4-2 *Perseverance*
(Photographed by Yang Fashun)

案例三：君子之交（图4-3）

　　作者：刘金亮

　　主要花材：梅、兰、竹、菊。

　　作者创意：采用梅、兰、竹、菊，以物喻人。人与人的交往应该是心与心的交换，表现的应该是高尚的情操与品德。

　　点评：梅、兰、竹、菊号称花中"四君子"，该作品以清雅、隽永、比例得当的构图，令4种花材穿插自然，表现出梅的坚韧挺拔、兰的飘逸幽香、竹的刚毅俊秀和菊的风霜不惧。

Case Three: Friendship between Gentlemen (Picture 4-3)

Arranger: Liu Jinliang

Main floral materials: *Armeniaca mume*, *Cymbidium* spp., Bambusoideae, *Dendranthema mordifolium*.

Originality: The work compares the four plants — *Armeniaca mume*, *Cymbidium* spp., Bambusoideae, *Dendranthema morifolium* — to human, conveying the idea that the communication between people should be the heart-to-heart exchange, with the expression of lofty sentiments and virtues.

Comments: In China, *Armeniaca mume*, *Cymbidium* spp., Bambusoideae, *Dendranthema morifolium* are known as the "Four Gentlemen" in plants. With an elegant, meaningful and proportionate composition, the work makes a natural arrangement of the four floral materials, and manifests the firmness and straightness of *Armeniaca mume*, the grace and fragrance of *Cymbidium* spp., the fortitude and handsome of Bambusoideae, and *Dendranthema morifolium*'s fearlessness of wind and frost.

图4-3 《君子之交》（杨发顺摄）
Figure4-3 *friendship between gentlemen* (Photographed by Yang Fashun)

案例四：天音（图4-4）

作者：郑丽

主要花材：牵牛、向日葵、狗尾草、百合、葱兰。

作者创意：玉兔西隐星光淡，日出东山好凭栏。惊觉两鬓清风荡，却是苍穹放歌来。

点评：作品花材简洁、朴实，向日葵代表初升的太阳，葱兰象征西沉的月亮，牵牛是唱歌的小喇叭，飘逸的狗尾草是天音的旋律。创作的诗与插花融合表达出在日起月落、宁静爽朗的清晨，人与自然神交、聆听自然声音的意境。绽放在琴弦上的郊野草花寓意着每个生命都有灵魂与歌喉，这是大自然通过一草一木奏出的天音。

Case Four: Music of Heaven (Picture 4-4)

Arranger: Zheng Li

Main floral materials: *Pharbifis nil* (Hmorning glory), *Helianthus annuus*, *Setaria viridis* (bristlegrass), *Lilium* spp., *Zephyranthes candida* (zephyranthes candida).

Originality: The moon set with dim starlight, and it was a good time to stand by the railing and overlook the scenery at the sunrise. I suddenly felt something passing my ears like a refreshing breeze, and it turned out to be the song from the heaven.

Comments: The materials used in the arrangement are simple and plain, with a *Helianthus annuus* standing for the rising sun, *Zephyranthes candida* for the setting moon, *Pharbifis nil* for singing trumpets, and flowing *Setaria viridis* for the melody of the sounds of heaven. The fusion of the poem written by the arranger and the flower arrangement expresses the conception of a person's spiritual communication with nature and listening to the sounds of nature in a peaceful and refreshing morning. The wild flowers bursting into bloom on the strings imply that every life has its soul and voice, and it is the music of heaven played by nature via every flower and grass.

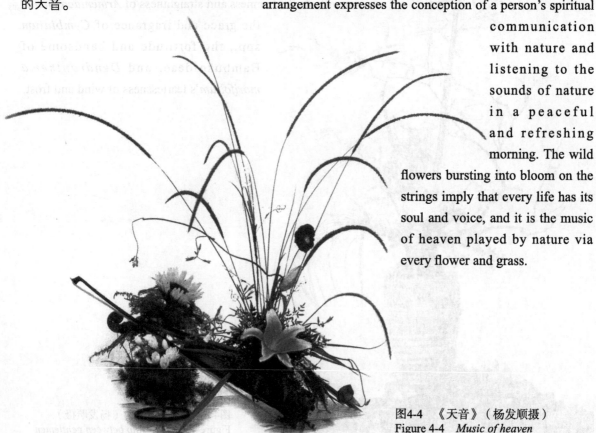

图4-4 《天音》（杨发顺摄）
Figure 4-4 *Music of heaven*
(Photographed by Yang Fashun)

案例五：喜见恩师（图4-5）

　　作者：郑丽

　　主要花材：鹤望兰、勿忘我、月季、菊花、情人草、火龙珠。

　　作者创意：昭昭日月二十载，通讯未传江山碍。一朝听闻恩师唤，中彩难及此心怀。六千学子勤灌溉，十年树木盼成才。二人相扶园丁路，班班育出秀苗来。

　　点评：该作品为喜见阔别20年的恩师而作，作者所配的藏头诗点明了学校和班级。老师与师母都是中学教师，两枝鹤望兰形似老师在翘首回望远方的学子，情人草是老师与师母情深意笃的见证；通过月季勾勒的"问号"象征恩师在教育的时候循循善诱、启发开导；"问号"中心红色的火龙珠心形果实表达了老师在教育岗位上呕心沥血、诲人不倦；菊花则是老师高风亮节的写照；勿忘我点明了学子对师恩的铭记。与其说这是一个插花作品，不如说是作者喜见恩师的思想和情感。

Case Five: A Happy Reunion with the Teacher (Figure 4-5)

Arranger: Zheng Li

Main floral materials: *Strelitzia reginae*, *Myosotis silvatica*, *Rosa chinensis*, *Dendranthema morifolium*, *Codariocalyx motorius*, *Hypericum*.

Originality: It has been 20 years since we parted, and the mountains and rivers blocked our association. When I heard the calling from my teacher, I was much happier than winning a lottery. My teacher has taught thousands of students, expecting them to achieve success. Supporting each other in career and life, the teacher couple has cultivated many talents.

Comments: The work is for the reunion of the arranger with her teacher after 20 years' separation. The Chinese version of the poem attached to the flower arrangement is an acrostic, which uses the first character of each verse to tell the school and the class in which she had studied. The teacher and his wife are both teachers, and the two *Strelitzia reginae* are just like the two who are expecting their students far away. *Codariocalyx motorius* is also called lover grass, and it stands for the couple's affectionate love. The "question mark" formed by *Rosa chinensis* symbolizes the teachers, guidance and enlightenment, and the red heart-shaped *Hypericum* cherry fruits in the middle of the "question mark" expresses the teachers' hearty efforts and tirelessness in teaching. The *Dendranthema morifolium* is a portrayal of the teachers' exemplary conduct and nobility of character, while the *Myosotis silvatica* shows that the students will never forget their teachers. The work is of the arranger's thoughts and feelings upon the reunion with her teacher rather than a flower arrangement.

第4章 插花命题及花文化内涵

图4-5 《喜见恩师》（郑丽摄）
Figure 4-5 *A happy reunion with the teacher* (Photographed by Zheng Li)

案例六：千古（图4-6）

作者：郑丽、包晓鹏、马龑、周文倩、金韬、高飞、柯燚

作者创意：为人正直品德崇高博学多能，昔时为师美德长与天地齐，光明磊落尽职尽责勤俭廉朴，今日永别英灵常存宇宙间。

点评：这是一组告别仪式的祭奠摆花，作品突破了普通殡仪用花的花圈花篮等造型，用简单的白色菊花插作立体的文字和梅花图案，置于往生者灵前。菊花花材、梅花图案与字面含义表达了对联的内容，而"千古"二字更是通过简单的笔画表现往生者的正直、磊落、俭朴。这便是"形"与"意"的结合。

Case Six: Immortality (Figure 4-6)

Arranger: Zheng Li, Bao Xiaopeng, Ma Yan, Zhou Wenqian, Jin Tao, Gao Fei, Ke Yi

Originality: Being upright and honest, noble and learned, you are immortal with heaven and earth, my dear teacher. Being aboveboard and unfeigned, responsible and industrious, you still live in our heart forever, though a farewell has to be made.

Comments: This is a set of flower arrangements for a farewell ceremony. Instead of the wreaths and flower baskets commonly used in funerals, this set of works uses simple white *Dendranthema morifolium* to form stereoscopic Chinese characters and patterns of *Armeniaca mume*, placed in front of the deceased. Those *Dendranthema morifolium*, *Armeniaca mume* patterns and characters convey the content of the couplet, while the topic "Immortality" expresses the uprightness, honesty and thrift of the deceased in two simple Chinese characters. It is exactly the combination of "form" and "meaning".

图4-6 《千古》（艾万峰摄）
Figure 4-6 *Immortality* (Photographed by Ai Wanfeng)

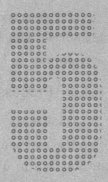

第 5 章　插花制作基础及常用手法

Chapter 5　Fundamentals and Commonly Used Techniques of Flower Arrangement

5.1 花艺制作技法概述

插花的过程就是艺术的创作过程，当今花艺制作中有了许多新技法。

5.1.1 花艺材料加工

（1）编织

运用柔软、有一定韧性、宽窄不同的长叶片或粗细不同的枝条，编织创作美丽的造型，从而丰富花艺作品。

（2）串联

将花瓣、叶片、花朵、果实等进行串联，形成不同的造型，丰富花艺作品，如图5-1所示。

（3）缠绕

缠绕是围绕一个中心旋绕，主要体现线条的装饰性，所以缠绕可在多枝花梗上进行，也可在一根花材的茎干上盘绕，甚至还可以围绕一个无形的中心进行，旨在表现线条的优美、流畅。

（4）裁切

想要创作出优美的造型，对花材的裁切是必需的，也是最基本的技法。

（5）粘贴

用植物材料粘贴成一个面或体，使原本单调的花器、桌面、背板等物体的表面产生肌理，成为主体花艺很别致的陪衬，这就是花艺设计中的粘贴手法。在现代花艺设计中，粘贴手法也越来越多地用于制作花艺作品的主体而成为观赏焦点。

图5-1 串联

5.1 General Introduction to Skills of Flower Arrangement

The process of arranging flowers is the creation of art. Many new skills have been used to enrich the flower art arrangement.

5.1.1 Processing of Floral Materials

(1) Weaving

Use soft and flexible long blade of different widths or branches of different thickness, and make beautiful modeling for weaving, so as to enrich flower artistic works.

(2) Connecting

Connect petals, leaves, flowers, fruits in series to form different shapes so as to enrich flower artistic works, as shown in Figure 5-1.

(3) Winding

Winding is to rotate around a center, mainly presenting decorative lines. So winding can be conducted either on many peduncles or on the stem of a flower. It can even be conducted around an invisible center with the purpose of showing the grace and fluency of lines.

(4) Cutting

In order to create graceful shapes, it is necessary to do some cutting work to floral materials and it is also the most basic skill.

(5) Pasting

Pasting is to produce a side or cube with plant materials, making originally monotonous floral vessels, desktops and backplanes emerge texture, and become chic foil for the main body of a floriculture work. Pasting skill in modern floral design is increasingly being used to make the main body of flower works and becomes the focus of appreciation. Dry

Figure 5-1 Connecting

如干燥叶片、干叶脉、枯枝、藤条等，一般用热胶粘贴。而新鲜花材如花瓣、花梗、叶片等用专用的鲜花胶粘贴。

（6）捆绑

捆绑是将线形材料围绕一个中心旋绕，强调功能上的要求，使作品更牢固。

5.1.2　花艺结构构成

花艺创作是运用各种花艺技巧，完成形式新颖的现代花艺造型，其结构构成有以下几种。

（1）架构

架构就是由不同形式的框架构成观赏的主体，再以鲜花装饰的现代花艺造型。架构式花艺作品比传统插花更具有层次感、空间感和立体感，同时还具有气度感、华丽感和时代感，目前已被广泛应用于各种花艺布置场合。架构是设计师采用联想、象征、意向等方法创建一个主题构架，以增强作品的立体感、层次感以及内部结构张力。其造型不受限制，构图更加自由，表现方法新颖独特，艺术魅力独具特色，符合现代人的审美情趣。

（2）堆叠

堆叠是指用同种或不同种花材堆叠起来形成一定的造型，以增加作品的厚度、质感等。堆叠的过程中，花与花之间没有空隙，如图5-2所示。

（3）平行线

平行线有直线平行和曲线平行，给人整齐、美观、协调的感觉，如图5-3所示。

（4）组群

将同一种颜色、类、种集中于一体，构成一个个集群，当多个集群

leaves, dry veins, deadwood and cane are generally stuck with hot glue, while fresh floral materials such as petals, peduncles and blades are stuck with flower glue.

(6) Binding

Binding is to rotate linear materials around a center with the emphasis on functional demands so as to make the work more solid.

5.1.2　The Structure of Floriculture

Artistic floriculture is to use a variety of art skills to create novel modern flower shapes, which includes the following types.

(1) Framing

Erecting is to form the main body of watching of a work with different forms of frameworks, and shape modern floriculture decorated by fresh flowers. Framework flower artistic works have more sense of layers, space and stereoscopy than traditional flower arrangement. It also has a spectacular feeling, luxuriant feeling and times feeling, which has been widely applied to a variety of floral arrangement. Framing is that designers construct a theme framework via association, symbolization, intention and the like to enhance the sense of stereoscopy, layer and the interior structure tension of the work. Because of the unstrained modeling, free composition, novel representing ways, it has a particular artistic charm and fits the aesthetic taste of people.

(2) Stacking

Stack the same or different floral materials together to form a certain shape so as to increase the thickness and texture feeling of the works. There are no gaps between flowers during the stacking process, as shown in Figure 5-2.

(3) Parallel lines

Parallel lines include linear parallelism and curve parallelism which give a person the feeling of tidiness, beauty and harmony, as shown in Figure 5-3.

(4) Grouping

Combine the same color, class and breed together to form a cluster. When multiple clusters constitute a body, it

图5-2 堆 叠
Figure 5-2　Stacking

图5-3 平行线
Figure 5-3　Parallel lines

构成一体时就形成组群。通过组群可以达到将同类、同色、相同类型、相同质感等的材料甚至异质素材组合一体，给人留下深刻印象。多运用于大体量花卉材料组合设计，如图5-4所示。

（5）支架

支架不作为作品观赏点，或者作品完成后完全看不到支架，只是便于放置花材。仅仅起造型、支撑的作用，如图5-5所示。

（6）交叉线

利用交叉线营造出放射效果，是不具中心点的任意直线交叉的杰作。最后加入一些交叉线条，提升空间占有率，从而产生适合大空间陈设的效果，如图5-6所示。

forms a group. The same class, color, type and texture, and even heterogeneous materials can be combined into a whole by the group, impressing people deeply. It is most applied to huge-scale floral material combination design, as shown in Figure 5-4.

(5) Scaffolding

Scaffolding doesn't work as a point of appreciation, or is invisible after the completion of the works. It is used only to place flowers with the role of modeling and supporting, as shown in Figure 5-5.

(6) Crossing lines

Using crossing lines to create a radiation effect is a skill for making a masterpiece which has arbitrary linear crossover but has no center point. At the end, some intersecting lines are added to promote space occupancy to make the work suitable for large space display, as shown in Figure 5-6.

图5-4 点、线、面、体综合立体构成
Figure 5-4　Comprehensive constitution of points, lines, surfaces and solids

图5-5 支 架
Figure5-5　Scaffolding

图5-6 交叉线
Figure 5-6　Crossing lines

5.2 常用表现手法

5.2.1 立体构成法

立体构成指的是三维空间。花艺作品以实际厚度、高度和宽度体现出立体造型。

（1）立体构成特点

①轮廓的不固定性：人在欣赏立体造型的全貌时会造成视点的移动，因而形成轮廓的不固定性。

②触觉感：立体造型虽然是静止的，但组成其造型的点、线、面在空间体现为立体而有动感，即有触觉感。

③光影感：立体造型一般都是通过明暗和落影来显示自身的体积效果，因而具有光影感。

（2）立体构成要素及基本形态

①立体构成要素：立体构成要素有形态要素、材料要素、形式要素等。

②立体构成基本形态：立体构成的基本形态有点、线、面、体。

（3）立体构成在插花艺术中的表现和运用

①线材的立体构成：直线与曲线是构成线的两大系统，也是决定一切由线构成的形的基本要素。一般直线表示静，曲线表示动。线材大致可分为软质线材（又称拉力材）和硬质线材（又称压缩材）两大类。软质线材包括棉、麻、丝、绳、化纤等软线，还有铁、钢、铝丝等可弯曲变形的金属线材；硬质线材有木、塑料及其他金属条材等，见图5-1。

5.2 Commonly Used Techniques of Expression

5.2.1 Three-dimensional Composition Technique

Three-dimensional composition refers to the three-dimensional space. Artistic floricultural works embody stereoscopic shape via their actual thickness, height and width.

(1) Characteristics of Three-dimensional Composition

① Irregularity of outlines: When a person appreciates the full perspective of a three-dimensional shape, his point of sight is mobile so that the outline becomes irregular.

② Sense of touch: Three-dimensional shape is static, but its points, lines and surfaces are three-dimensional and dynamic in the space, namely having the sense of touch.

③ Sense of light and shadow: Three-dimensional shape displays its own size effect through light and shade, and falling shadows, which consequently has a sense of light and shadow.

(2) Components and Fundamental Forms of Three-dimensional Composition

① Components of three-dimensional composition: The components of three-dimensional composition elements include form, material, shape and etc.

② Fundamental forms of three-dimensional composition: The fundamental forms of three-dimensional composition include points, lines, surfaces and solids.

(3) Manifestation and Application of Three-dimensional Composition in Artistic Flower Arrangement

① Three-dimensional composition in lines: Straight lines and curve lines are two big systems of lines, which are the basic elements of the shapes composed of lines. Generally linear lines represent static state while curve lines represent dynamic state. Lines can be roughly divided into soft lines (also tension lines) and hard lines (also compression lines). Soft lines include cotton, hemp, silk, rope, chemical fiber and bending metal lines such as iron,

②面材的立体构成：面材是由长、宽二维空间的素材构成的立体造型。面材具有平薄和扩延感，如图5-7所示。

③体的立体构成：体的立体构成是以三维的有重量、体积的形态在空间构成完全封闭的立体。占据三维空间，可以产生较强烈的空间感；相对于点立体、线立体和面立体更具重量感、充实感；能产生稳重、秩序、永恒的视觉感受；不规则的体能产生亲切、自然、温馨的感觉，如图5-8所示。

④点、线、面、体综合立体构成：空间的立体构成是由点、线、面、块占据或围合而成的三度虚体，具有形状、大小、材质、色彩、肌理等视觉要素，以及位置、方向、重心等关系要素，其效果也深受这些要素的影响，见图5-4。

steel, aluminium wire, and etc. Hard lines include wood, plastic and other metal materials, as shown in Picture 5-1.

② Three-dimensional composition of surfaces: Surfaces are three-dimensional composition composed of materials with the two-dimensional space of length and width. Panels have the sense of flat and expansion, as shown in Figure 5-7.

③ Three-dimensional composition of solids: Solids is composed of three-dimensional shape with weight and volume in space, which is completely closed. Occupying the three-dimensional space can produce a strong sense of space. Compared with the points stereoscopy, lines stereoscopy and surfaces stereoscopy, solids stereoscopy has more sense of weight and enrichment. It has the visual perception of being steady, orderly and everlasting. Irregular solid has the sense of kindness, ease and warm-heartedness, as shown in Figure 5-8.

④ Comprehensive three-dimensional composition of points, lines, surfaces and solids: Three-dimensional composition of space is a three-dimensional insubstantiality formed by points, lines, surfaces and solids. It has visual elements of shape, size, material, color, texture and relational elements such as location, direction, and core. The effect is also influenced by these factors, as shown in Figure 5-5.

图5-7　面材的立体构成
Figure 5-7　Three-dimensional composition of panels

图5-8　体的立体构成
Figure 5-8　Three-dimensional composition of body

5.2.2 一枝突出法

一枝突出法是突出不同种类的花、叶、枝,加以夸张变化的手法,达到特定环境空间所需要的装饰效果。如让枝条保持自然的生长态势,仿佛从瓶中破土而出,生机盎然。

(1)一枝突出法在插花艺术中的表现和运用

①表现个体花材的特质,如图5-9所示;

②使花材布局聚中有散、静中有动,如图5-10所示。

(2)一枝突出法在插花中的类型

①枝的一枝突出,如图5-11所示;

②叶的一枝突出,如图5-12所示;

图5-10 花材布局聚中有散、静中有动
Figure 5-10 Creating the floral arrangement of scattering mingled with gathering and the static mingled with the dynamic

5.2.2 Highlighting-One-Branch Technique

Highlighting-One-Branch Technique is an exaggerated method which highlights the different categories of flower, leaf and branch so as to achieve adornment effect necessary for specific environment. For example, keeping branches natural growth momentum looks as if it breaks out of the bottle and shows exuberant.

(1) Manifestation and application of highlighting-one-branch technique in artistic flower arrangement

① Demonstrating the nature of a particular floral material, as shown in Figure 5-9.

② Balancing scattering and gathering, static and dynamic in arrangement, as shown in Figure 5-10.

(2) Types of highlighting-one-branch technique in artistic flower arrangement

① Highlighting-one-branch in branch arrangement, as shown in Figure 5-11.

② Highlighting-one-branch in leaf arrangement, as shown in Figure 5-12.

图5-9 表现个体花材的特质
Figure 5-9 Demonstrating the nature of a particular floral material

图5-11 枝的一枝突出
Figure 5-11 Highlighting-one-branch in branch arrangement

图5-12 叶的一枝突出
Figure 5-12 Highlighting-one-branch in leaf arrangement

图5-13 花的一枝突出
Figure 5-13 Highlighting-one-branch in flower arrangement

③花的一枝突出，如图 5-13 所示。

5.2.3 斜角呼应法

（1）斜角呼应法在插花艺术中的表现和运用

①平中出奇，如图 5-14 所示；

②加强动感，如图 5-3 所示；

③对角平衡，如图 5-15 所示。

（2）斜角呼应法在插花中的类型

①造型的斜角呼应，如图 5-16 所示；

②色彩的斜角呼应，如图 5-17 所示。

③ Highlighting-one-branch in flower arrangement, as shown in Figure 5-13.

5.2.3 Oblique Angle Echoing Technique

(1) Manifestation and Application of Oblique Angle Echoing technique in Artistic Flower Arrangement

① The extraordinary out of the ordinary, as shown in Figure 5-14.

② strengthening dynamic, as shown in Figure 5-3.

③ Balance of opposite angles, as shown in Figure 5-15.

(2) Types of Oblique Angle Echoing technique in Artistic Flower Arrangement

① Oblique angle echoing in shaping, as shown in Figure 5-16.

② Oblique angle echoing in color, as shown in Figure 5-17.

图5-14 平中出奇
Figure 5-14 The extraordinary out of the ordinary

图5-15 对角平衡
Figure 5-15 The balance of opposite angles

图5-16 造型的斜角呼应
Figure 5-16 Oblique angle echoing in shaping

图5-17 色彩的斜角呼应
Figure 5-17 Oblique angle echoing in color

5.2.4 分组组合法

分组组合法就是将形状、颜色相似的花朵进行分组、分区，每一组之间都需要有一定的距离。

（1）分组组合法在插花艺术中的表现和运用

①组群布置；

5.2.4 Grouping Technique

Grouping technique is to group and partition flowers with similar shape and color. There is a certain distance between each group.

(1) Manifestation and Application of Grouping technique in Artistic Flower Arrangement

① Grouping arrangement.

②组群形式，如图 5-18 所示；

③花材多而不杂、变而不乱，如图 5-19 所示。

（2）分组组合法在插花中的类型

①按花材不同品种分组组合，如图 5-20 所示；

②按花材不同色彩分组组合，如图 5-21 所示。

（3）注意事项

①注意各组花材的分割与联系；

②有些花材可以散插。

② Forms of grouping, as shown in Figure 5-18.

③ Floral materials being rich but not mussy, and changeable but not messy, as shown in Figure 5-19.

(2) Types of Grouping technique in Artistic Flower Arrangement

① Grouping according to types of floral materials, as shown in Figure 5-20.

② Grouping according to colors of floral materials, as shown in Figure 5-21.

(3) Dos and Don'ts

① Paying attention to the separation and connection of groups of flowers.

② Scattering some floral materials.

图5-18 组群形式
Figure 5-18 Forms of grouping

图5-19 花材多而不杂、变而不乱
Figure 5-19 Floral materials being rich but not mussy, and changeable but not Messy

图5-20 按花材不同品种分组组合
Figure 5-20 Grouping according to types of floral materials

图5-21 按花材不同色彩分组组合
Figure 5-21 Grouping according to colors of floral materials

5.3 插花基本技巧

5.3.1 准备工作

准备工作从选材开始，选材一般遵循以下几方面。

（1）选材要点

①东方式插花：因东方式插花主要由三主枝构成，所以对于主枝的形态极为讲究。通常选取形态富于变化、具有线条美感和韵律的树枝或藤条。

②西方式插花：西方式插花的选材关键是花材的质量和色彩的搭配。要求花材种类不能太多，花朵大小尽量一致，体态丰满，色彩搭配协调。

（2）优质花材的选择：

①保鲜时间长，耐水养，不易凋萎。

②生长良好，花蕾大，花梗壮且挺拔。茎、叶、花朵有生气、光泽，无疤痕、虫眼、病态。

③具有一定观赏价值。

5.3.2 修理花材

（1）一般整理性修剪

①去除所有枯萎、有虫、有病、受伤的枝、叶、花朵。

②去除花材下部多余的叶片。

③将花材下部切口处斜剪约4cm。在气候干燥季节或对于一些吸水性较差的花材，应当在水中剪切。

④去刺。如月季等花材有刺，宜使用前去刺，可用玫瑰钳或小刀削除。

5.3 Basic Skills of Flower Arrangement

5.3.1 Preparation

The preparation starts with the selection of materials, which generally follows the following aspects.

(1) Gist of material selection

① Eastern flower arrangement: The eastern flower arrangement mainly consists of the frame of three major branches, so branch form is important. In general, branch or cane that are rich in forms and with the beauty of lines and lingering charm are will be selected.

② Western flower arrangement: The western flower arrangement puts emphasis on the quality of flowers and color collocation. Greater variability of flower species is not advocated. Try to select flowers that are of similar sizes, full in shape and with harmonious color collocation.

(2) Selection of High-Quality Floral Materials

① Long duration of freshness, water-tolerance, not readily withering away.

② Well-grown, big bud, thick and straight peduncle, lustrous stem, leaf and petal without scar, worm sting and morbidness.

③ Being of ornamental value.

5.3.2 Trimming of Floral Materials

(1) General Trimming

① Remove all withered, defective, sick and wounded branches, leaves and flowers.

② Remove unnecessary blade in the lower part of flowers.

③ Diagonally cut around 4 cm in the lower part of the flowers. This step should be carried out in water under the circumstances when dry seasons come in some flowers have poor water-absorption capacity.

④ Remove thorn. For example, before arrangement, rose pliers or knives can be used to cut the thorns of *Rosa chinensis*.

（2）造型修剪

①花朵的处理与造型：

剥花、剥芽：剥花主要指剥除局部花瓣或花朵。如剥掉月季的最外几层花瓣，使花朵显得新鲜、整齐、含蓄；去掉唐菖蒲、满天星等先开的凋谢小花；鹤望兰、黄金鸟等隐藏在佛焰苞内的小花和香石竹尚未开放的花朵，可用手轻轻拨一下，使其开放角度增大；睡莲有日开夜合的习性，插花前应先将萼片剥去，以保证花的开放。

落瓣：用人工除去花材局部或全部花瓣，改变花材的大小，并创造一种残缺的美感。如花盘过大的向日葵，可将外围花瓣全部去掉；重瓣非洲菊可采用外圈花瓣局部落瓣的方法处理。半落瓣最经典的应用是荷花，荷花有非常多的变化，可表现花开花谢的效果。马蹄莲的漏斗状佛焰苞凋萎时，全部落瓣后只留存肉穗花序，也具有一定的观赏价值。

去花药：在百合初放期，应及早摘除雄蕊上的花药。因为百合成熟时，花药会裂开，释放大量花粉，不仅容易玷污花瓣，影响花朵纯洁的外形，还会粘染衣服，不易清除。

补强：主要用于花枝易折断或花茎细软的花材（睡莲、水仙等），而重瓣、花大的非洲菊、牡丹等，因其花朵硕大，也可采用金属丝扶持补强，以免垂头。

②枝的处理与造型：

铁丝缠绕校形：用铁丝螺旋状缠绕于茎、梗上，然后用手慢慢将其扳弯。

(2) Shaping Trimming

① Treatment and shaping of flower:

Stripping and debudding: Stripping means peeling off partial petals, such as the outer petals of *Rosa chinensis*, this treatment will make the flower appear fresh, neat and subtle. Remove the withered petals of *Gladiolus gandavensis* and *Gypsophila paniculata*. For the hidden petals on *Strelitzia reginae* and *Caragana sinica*(heliconia) and the petals of *Dianthus caryophyllus* that have not bloomed yet, you can slightly put them apart by hand to allow a wider blooming. *Nelumbo nucifera*is likely to bloom in the day and tuck at night, so you should strip off the sepals in advance to ensure that they will bloom.

Remove petals: It means artificially removing off partial or all petals of flowers to make them look smaller, thus creating an atmosphere of incomplete beauty. For example, all of the outer petals can be removed from the *Helianthus annuus* with excessively large discs. Outer petals can be partially removed from double *Gerbera jamesonii*. Half-fallen petals are most common in *Nelumbo nucifera*. When the funneled spathe of *Zantedeschia aethiopica* lilies withers away and falls off, only spadix will remain. And this is what the ornamental value lies in.

Remove anther: Removing the anther from stamen as soon as possible once *Lilium* spp. begins to bloom because anther is likely to crack and release a great deal of pollen when it is mature. If not properly handled, the anther will tarnish the petals, affect the pure appearance of flower and stain cloth.

Reinforcement of flower frame: It is mainly applied to those flowers, such as *Nelumbo tetragona* and *Narcissus tazetta*, whose branch breaks easily or the stem is thin and soft. For double *Gerbera jamesonii* Bolusand *Paeonia suffruticosa* with huge petals, a piece of wire can be used to support them and prevent them from drooping.

② Treatment and shaping of branch:

Shaping with iron wires: Twine spiral wire on the stem and peduncle, and then bend them slowly by hand.

铁丝穿心：花葶中空的非洲菊、花梗组织疏松的马蹄莲等草本花材，可用粗细适中的铁丝从茎基部中央由下往上穿进去，或从花蕊处由上而下插入，将其扳成所需形状。

枝条弯曲：花材的粗细、软硬及易弯曲程度不同，使用的方法不同。粗大的树干可用锯或刀先锯1~2个缺口，深度为枝粗的1/3~1/2，嵌入小楔子，强制弯曲。如枝条较脆易断，可将弯曲的部位浸入热水中（也可加醋），取出后立刻放入冷水中弄弯。花叶较多的树枝，须先把花叶包扎遮掩好，直接放在火上烤，每次烤2~3min，直到树枝柔软、足以弯曲成所需角度为止，然后立即放入冷水中定型。枝条较软的如连翘等，用两个拇指对放在需要弯曲的位置，慢慢扳动枝条即可。

③叶的处理与造型：

修剪：如将棕榈或蒲葵叶片剪成船帆形、蝶形、飞雁形；将针葵或散尾葵叶片剪成锯齿形、塔形、扇形；将八角金盘叶片剪成梅花形、圆形、扇形等。

弯曲：如在巴西铁背后用胶带贴上两根铁丝，可以弯曲成翻滚的波浪；将一叶兰叶片顶端卷成圆筒等。

弯折：如水蜡烛、剑叶龙血树等，在叶片的某一点，作折而不断的弯折处理，可折一次或几次，几次弯折可形成一定造型。

Stick iron wires into flowers: For herbaceous flowers like *Gerbera jamesonii* which is hollow in scape and *Zantedeschia aethiopic* with loose peduncles, iron wire of appropriate thickness can be used to stick into flowers through the stem in a bottom-up way or through the stamen in a top-down way and then you can pull it into the intended shape.

Bend branch: The methods vary with the degree of thickness, hardness and flexibility of flower materials. In the case of thick branch, saws or knives can be used to saw one or two notches, the depth of which is 1/3 to 1/2 the thickness of the branch. And then you can embed a small wedge into the notch for forced bending. For those brittle branches, the flexuous part can be immersed in hot water (or with the addition of vinegar). Next, you should take it out immediately and put it into cold water for bending. As to those branches with numerous floral leaves, after wrapping the leaves well, you can directly place them above the fire for 2 to 3 minutes until the branches are soft enough to bend into the intended angle and then put them immediately into cold water for shaping. Soft branches like *Forsythia suspensa* just need the placement of two thumbs in the appropriate position to slowly pull them.

③ Treatment and shaping of leaf:

Pruning: The leaves of *Trachycarpus fortunei* and *Livistona chinensis*(livistona chinensis) can be pruned into the shapes of sails, butterflies and wild gooses; that of *Phoenix loureirii*(roebelenii) and *Chrysalidocarpus lutescens* (chrysalidocarpus) can be trimmed into the shapes of zigzag, tower and fan and that of *Fatsia japonica* (fatsia) into the shapes of plum blossom, circle and fan.

Bending: The effect of rolling wave will be produced if you use tape to attach two iron wires to the back of *Dralaena fragrans*. The top side of *Aspidistra elatior* leaf can be rolled into the shape of cylinder.

Folding: The leaves of *Dysophylla yatabeana*(typha latifolia) and *Dracaena cochinchinensis* can be folded at any point of the blade once or several times but not be broken off. Repeated folding will contribute to the shaping of the leaf.

叶片撕裂：平行脉的叶片可顺着叶脉将叶片尖端撕成几条，或将叶片从中间撕开，或沿叶脉纹路撕去部分叶面。

圈叶：适用于有长柄或中脉较坚韧的长条形叶片，可先在叶片的尖端开一小孔，将叶片弯曲，令叶柄穿入小孔，使叶的线条闭合成圆弧，如一叶兰、巴西铁等。

翻翘：适用于长条形和有平行脉的叶片，如麦冬、一叶兰、剑叶龙血树等。先在叶片的中部纵开一小口，将叶子的前端从开口处穿入、拉过。

打结：选择细长的叶材，如麦冬、剑叶龙血树、巴西铁等。

编织：将散尾葵、针葵等羽裂叶片像编席子般编织成扇形或牛角形。

5.3.3 花材固定

（1）花泥固定

使用花泥固定花材时，先将吸足水的花泥按花器的大小切成块，花泥应高出容器口3cm左右，其厚度可视花型需要及下垂枝干的角度而定。为稳固花泥，可用防水胶带将花泥固定在花器上。当花器较深时，可在花泥下面放置填充物。若花器为编制的花篮不能盛水，可在花泥下面包垫塑料纸，以防花泥漏水。口径较大的容器插粗茎花材时，可用铁丝网罩在花泥外面，以增强支撑力。花枝插入花泥的深度为3cm左右。使用花泥固定花材时，应在插前仔细观察构思，确定好要插的位置与角度，争取一次插置好，尽量避免反复。

Tearing: The leaf with parallel vein can be torn into several stripes along the veins from the top end. You can also tear apart the center of the leaves or tear the leaf apart along the line of vein.

Circling: It is applied to those of long and narrow leaves with long handle or hard midrib. You can cut a small hole at top end of the leaf and then bend it to put petiole through the hole, making the line of leaf, such as pleione leaf and *Dralaena fragrans* form a circular arc.

Warping: It is applicable to long and narrow leaves with parallel vein, such as *Ophiopogon japonicus*, *Aspidistra elatior* and *Dracaena cochinchinensis*. You can firstly cut a small longitudinal opening in the middle of the leaves and then pull the front-end of the leaves through the opening.

Knotting: Select vimineous leaf, such as *Ophiopogon japonicus*, *Dracaena cochinchinensis* and *Dralaena fragrans*.

Weaving: Similar to the weaving of mats, the pinnatifid leaves like *Chrysalidocarpus lutescens* and *Phoenix loureirii* can be woven into the shapes of fan or ox horn.

5.3.3 Fixing of Floral Materials

(1) Fixing with flower mud

Before using mud to fix flowers, you should cut the mud full of water absorption into pieces according to the volume of flower containers. Flower mud should be about 3cm higher than containers and the thickness depends on the shape of flowers and the angle of drooping branches. In order to fix flower mud, waterproof tapes can be used to fix mud to containers. Fillers can be placed under mud in deep containers. For woven baskets that cannot hold water, you can put a piece of plastic paper under flower mud to avoid water leakage. When choosing containers with large diameters arrange flowers with thick scape, you can cover the flower mud with wire meshes to strengthen support. The branches of flowers should be inserted about 3cm deep into the mud. Before fixing flowers, you should observe carefully and think about, determining the position and angle for arrangement. Strive to succeed in the first attempt so as not to redo it.

（2）剑山固定

用剑山插花前必须先向容器中加水，水位要高于剑山的针座，以便花枝插上后能吸水。一般花枝采用直接插入法。如果插直立枝条，将枝条基部剪平，直立方向插入即可。如果要做倾斜造型，应先斜剪，再直立插入，然后按构图需要将其压至所需的倾斜度。太细的花枝难以固定时，可在花枝基部绑缚一小段枝或套插于其他短茎中，或者将若干细小的花材扎成一小束再插，扩大花材与剑山的接触面。对于插较粗硬的木本花材，可将花枝基部竖向剪成十字形或一字形裂口，以利于枝条插入针座。

此外，瓶插时，还可采用瓶口内十字架或井字架固定、折枝固定、铁网丝固定等方法。

5.3.4 作品整理

作品完成后应注意检查花泥是否遮掩好，花器是否干净，摆放的位置是否整洁干净。同时要保持插作场地和周边环境的清洁。这是插花不可缺少的一环，也是插花者应有的素质。

(2) Fixing with receptacles

Add water first before arranging flowers in receptacles. The water level should be higher than the needle base of receptacles, making it easy for arranged flowers to absorb water. Ordinary flowers can be directly inserted into receptacles. For upright shaping, the base of branches should be cut flat and then vertically inserted. For oblique shaping, the base should be cut diagonally and vertically inserted and then press the branches to intended inclination. When some branches are too thin to be fixed, a small branch can be tied to the base of flowers or you can insert the branches into other short stems. It is also possible to bind several thin branches together because we can increase the contact area between flowers and receptacles by doing so. A vertical cross-shaped or line-styled notch can be cut at the base of flowers to insert the branches of arborescent flowers into needle bases.

In addition, placements of crosses, well character frames, folding branches and wire meshes in bottles are alternatives for bottled flowers.

5.3.4 Finishing Works

After the completion of the work, you should check whether flower mud has been covered well and whether they have been placed in clean and tidy areas or not. Clean up the place and keep it tidy. It is an indispensable part of flower arrangement and an essential quality for flower arrangers.

第6章 花材保鲜及包装

Chapter 6 Preservation and Packaging of Flowers Meterials

6.1 花材保鲜

6.1.1 花材萎蔫的原因

花材经采切，脱离了母体植株的营养和水分供应，生命活动受到干扰。在衰老过程中，细胞结构遭到破坏，内含的多种有机物、矿物质和水分等发生分解变化。造成花枝衰败和凋萎的具体原因有以下几点。

（1）水分供应入不敷出

植物体内水分代谢平衡是植物细胞维持正常代谢活动的基础。水分代谢是鲜切花采后的主要生理过程。鲜花中的水分含量一般占鲜花重量的80%以上，鲜花在剪切时，切断了水分供应，水分代谢失调，含水量下降，鲜花重量明显下降，导致各种生命活动异常，花朵萎蔫。鲜花衰败过程最明显的变化就是失水凋萎。

植物体主要通过输导组织的导管吸收水分，通过叶片及花朵的气孔散失水分。水分的吸收与散失之间只有保持平衡，才能使切花保持新鲜状态。水分的吸收和散失强度受多种因素制约，既有环境因素（如空气温度、湿度、空气流通性、光照强度等），也有切花自身的因素（如组织结构、叶片表面积等）。如果切花细胞排列紧密并含有较多的保护物质，如蜡质、角质层，或者气孔较少，散失强度就会小一些。但仅有少数切花自身有较好的保护性组织，如花烛的表面有厚厚的蜡质保护层。鲜花采切后，会不断散失水分，使鲜花体内水分的吸收和散失逐渐失去平衡。水分供应不足，导

6.1 Preservation of Flower Materials

6.1.1 Causes of Flower Materials Wilting

After being cut off and without nutrition and water supply from the mother plants, the living activities of flower materials are disturbed. In the process of their senescence, their cell structures are damaged, and there occurs the decomposition of water, many kinds of organic compounds and minerals in the cut flowers. The causes of cut flowers wilting are as follows.

(1) Water supply failing to satisfy the demand

Plants' water metabolism balance is the basis for plant cells to maintain normal metabolism. And water metabolism is also an essential physiologic process after flowers are cut off. On average, water content accounts for more than 80% of the weight of the flowers. Flowers are cut off, and so is its water supply. The water metabolism disorders, the decreasing water content and the significantly decreasing weight of cut flowers lead to all kinds of abnormal living activities and wilting flowers. In the process of wilting, the most striking change results from its dehydration.

The moisture absorption in plants is primarily carried out by plant vessels and the water loss is by the stomata of plant leaves and flowers. Cut flowers can not be kept in their fresh state unless the balance between the absorption and the loss of water is maintained. The absorption and the water loss are attributed to many factors, including both environmental factors (such as air temperature, humidity, air circulation and light intensity), and cut flowers themselves (such as the structure of tissues and leaf surface area). If cut flowers cells are closely arranged and contain comparatively more protective materials like wax and the cuticle, or less stoma, there will be less intensity of water loss. However, only a few of cut flowers themselves have better protective tissues, for example, there is a thick layer of protective wax on the surface of *Anthurium* spp. After cut flowers are cut off, they suffer from the continuous loss of water, and hence appears the imbalance between moisture

致花朵萎蔫或凋零。为减少水分散失，可及时摘除一些不必要的叶片。将鲜切花及时置于水中，也能够使之补充部分水分。

另外，在剪切花枝时，花枝剪口处往往会被挤压，造成损伤。空气从切口进入导管，会阻碍花枝吸收水分，同时剪口处受伤细胞产生的分泌物也会影响水分的输导。另外，水中常含有一些微生物，如细菌、真菌等，随着微生物的繁殖，菌丝体不断侵入，也使花枝在很大程度上失去了吸水能力。大多数切花对水分缺失十分敏感，并且也很容易失水。

（2）储藏养分消耗殆尽

鲜花要维持正常的生命活动，除了需要充足的水分，还需要充足的养分，以便通过呼吸作用，转化成维持生命活动的能量。碳水化合物与蛋白质是植物的重要组成成份，碳水化合物多以糖和淀粉的形式存在。糖是重要的呼吸基质；蛋白质是细胞的结构成分，是各种酶的组成部分。这些酶具有催化各种生理生化反应，使切花维持正常生命活动的作用。在鲜花衰败过程中，常伴随着蛋白质含量的减少，这是由于蛋白质分解加速而合成减少造成的。蛋白质是植物生命活动的物质基础，它的分解标志着生命活动的减弱。

然而，大多数人只是把鲜花插在水中，脱离了植株母体的鲜花，在室内养分消耗大于积累，一旦因呼吸作用将储藏的养分消耗殆尽，花朵必然凋萎。

absorption and water loss. The lack of timely water supply results in the wilting of flowers. In order to reduce water loss, it is advisable to remove some unnecessary leaves. In addition, placing cut flowers in water in time can also help supplement part of moisture.

In addition, in the process of cutting flowering shoots off, incisions tend to be crushed and left with crush injuries. From the incisions into the plant vessels, the air may prevent the moisture absorption of sprays. Meanwhile, the secretions produced by the injured cells on the incision will affect the moisture migration. Furthermore, water often contains some microorganisms, such as bacteria and fungi. With the reproduction of microorganisms and the continual mycelium invasion, the sprays lose their capacity of water absorption to a great extent. Most cut flowers are quite sensitive to water loss and are also easy to suffer from it.

(2) Nutrients in storage running out

Besides adequate water supply, sufficient nutrients are necessary for cut flowers to maintain normal living activities so that the nutrients can be converted into the energy to sustain living activities of cut flowers through respiration. Carbohydrates and proteins are important substances of plants. Carbohydrates exist in the form of sugar and starch, and sugar is an important respiratory matrix. Proteins are the structural components of cells and also compose a variety of enzymes, which can promote all kinds of physiological and biochemical reactions and which can help maintain cut flowers' normal living activities. The wilting of flowers is often accompanied by the decreasing protein content, because of the accelerating decomposition and the decreasing composition of proteins. Proteins function as the material basis of living activities of plants, so their decomposition marks the weakening living activities.

However, most people only immerse flowers in water. Away from their mother plants, flowers consume more nutrients indoors than what is accumulated. Once the nutrients in storage run out due to the respiratory action, flowers are doomed to wilt.

（3）产生有害物质，加速其衰败

鲜花体内进行正常生命活动时，会分解碳水化合物、蛋白质和脂肪，生成能量，同时也产生一些有害的物质。例如，多元酚是蛋白质的分解产物，它使许多草本切花的持久性降低。花枝本身还会产生乙烯，乙烯有促进花衰老的作用。乙烯对切花的影响决定于乙烯的浓度、作用时间及切花的敏感性。不同种类的切花对乙烯的敏感程度不同，香石竹、金鱼草、石斛兰、仙客来、水仙等切花都属敏感花卉，它们易受乙烯影响而衰败。菊花对乙烯不敏感，而月季、郁金香等敏感程度介于两者之间。即使是同一种花，也会因其发育程度不同而对乙烯的敏感性不同。一般情况下，蕾期对乙烯不太敏感，花盛开时对乙烯比较敏感。乙烯对切花的伤害症状各式各样，总的来说可归结为两大类型：一类是花冠蜷缩、褪色至凋萎，如香石竹、石斛兰；另一类表现为器官脱落，如月季、天竺葵的花瓣脱落，一品红的苞片脱落，金鱼草、香豌豆的整个小花脱落。

植物体的衰老器官（如衰老的花、叶片）是产生乙烯的主要器官。受伤或被病菌感染的枝也会产生乙烯，因此，应尽量避免切花染病或机械损伤。乙烯的形成还与温度、环境条件的氧气压有关。氧含量低及低温环境都能抑制乙烯的形成，因此，在鲜花采收后的各个处理环节，应创造低温、低氧的环境条件。总之，乙烯会加速花朵的衰败。蔬菜、水果中都含有乙烯，如果插花摆放

(3) Production of harmful substances fastening cut flower's decay

In the normal living activities, flowers will not only decompose carbohydrates, protein and fat, and generate energy, but also produce some harmful substances. Take polyhydric phenol for example. As the product of the process of protein decomposition, it reduces the durability of many herbal cut flowers. Besides, flowering shoots themselves will produce ethylene, which functions as an accelerator of flowers' wilting. Ethylene's effects on cut flowers depend on ethylene's concentration, action time, and cut flowers' sensitivity. Different kinds of cut flowers are sensitive to ethylene to a different degree. Cut flowers like *Dianthus caryophyllus*, *Antirrhinum majus, Dendrobium nobile, Cyclamen persicum* (cyclamen) and *Narcissus tazetta* var. *chinensis* are sensitive and easily affected by ethylene and thus wilt. *Dendranthema morifolium* is not sensitive to ethylene while the sensitivity of *Rosa chinensis* and *Tulipa gesneriana* to ethylene is between these two extremes. Even to the same kinds of cut flowers, their sensitivity to ethylene varies with their different degrees of growth. In general, flowers in bud are less sensitive to ethylene and flowers in full bloom are more sensitive. Ethylene causes various injury symptoms which can be divided into two types in general: the first type is that corollas of flowers like *Dianthus caryophyllus* L., *Dendrobium nobile* will roll up, fade and even wilt; the second type is that organs of flowers will come off, for example pedals of *Rosa chinensis* and *Pelargonium hortorum* (geranium), bracts of *Euphorbia pulcherrima* and florets of *Antirrhinum majus* and *Lathyrus odoratus* (sweet pea) will come off.

The declining plant organs such as aging flowers and leaves are the major organs to produce ethylene. Injured branches or branches infected by bacterium will also produce ethylene. Therefore, infection or mechanical damages to cut flowers should be avoided to the greatest extent. The formation of ethylene is also related to temperature and oxygen pressure in the environmental conditions. Low oxygen content and the environment with low temperature can inhibit the formation of

在它们附近,即使这种外源乙烯很微量,也会诱导和加速花材释放乙烯,导致花朵衰败。

(4)浸泡鲜花的水质不纯

水中的含盐量应适度,含氟量不宜太高。氟可使叶片气孔扩大,影响花材体内水分平衡,导致鲜花萎蔫。

(5)储运过程中温度过高

切花呼吸产生的能量大部分以热的形式散发出来,它对储运过程中的保鲜是有害的。无氧呼吸产生的酒精也是有害的。呼吸热的产生提高了切花的温度,对切花自身的呼吸具有促进作用,加速了切花体内储藏物质的消耗。它也有利于微生物的活动,加速切花的腐烂。因此,切花在储运过程中由于呼吸所造成的高温会加速鲜花的萎蔫,必须采取某些措施来延缓鲜花凋萎,延长观赏期。

6.1.2 花材的保鲜原理

将鲜切花存放于低温环境下或使用科学配制的保鲜剂等,可以延缓鲜花凋萎,其原理如下:

(1)存放于低温环境

鲜花在室温0~10℃比在20~30℃环境下的呼吸强度低很多,可以显著降低代谢活动,减少无益的自身消耗。因此,鲜切花存放于低温环境中是鲜花保鲜的一个重要措施。

ethylene. As a result, in each processing link after the harvests of flowers, it is advisable to create the environmental conditions with low temperature and low oxygen content. In short, ethylene will accelerate the wilting of flowers. Both vegetables and fruits contain ethylene, so if cut flowers are placed near them, it will induce and accelerate the release of ethylene of flowering shoots and hence resulting in the wilting of flower materials, even if this exogenous ethylene is of minimum amount.

(4) Impure soaking water

The quantity of salt should be moderate and fluorine content should not be too high. Fluoride can expand the leaf stoma, influence the balance of moisture of cut flower and finally result in the wilting of flowers.

(5) High temperature in the process of storage and transportation

Most energy produced in respiratory action of cut flowers is emitted in the form of heat, which is detrimental to flower preservation in the process of storage and transportation. Alcohol produced in anaerobic respiration is also harmful. Respiratory heat improves cut flowers' temperature, promotes their own respiration and accelerates the consumption of storage in cut flowers. Furthermore, it is beneficial to microbial activities and can accelerate cut flowers' decay. Therefore, high temperature in the process of storage and transportation can accelerate the wilting of cut flowers. It is necessary to take some measures to delay the wilting of flowers so as to extend their ornamental period.

6.1.2 Principles of Preserving Flower Materials

Storing in the environment with low temperature or applying scientifically-made preservative solutions can delay cut flowers' wilting. Its principles are as follows:

(1) Storing in low temperature

Flowers have much lower respiration intensity between 0℃ and 10℃ than they do between 20℃ and 30℃, which can reduce their metabolic activities significantly and their own unprofitable consumption. Therefore, one of the

（2）在保鲜液中放入杀菌剂

切花插入水中并不能完全保证水分的供应，这主要是由于其生理原因及微生物堵塞输导系统造成的。水中常含有一些微生物，如细菌、真菌，它们可以侵入导管，并分泌一些有害的代谢产物，影响鲜切花对水分的吸收。为此，在保鲜液中必须加入一些杀菌剂，以消除微生物的不利影响。

（3）在保鲜液中放入酸性物质

切花即使放置于无菌水中，没有微生物的毒害作用，也会有水分运输不畅现象发生，这是由于切花自身的生理原因造成的。在切花基部切口处，受伤细胞会产生分泌物，如一品红切花切口可以分泌白色乳状液体。部分切花在切口附近会产生一些氧化物等黏性物质，它们聚集在切口附近或水面上2~3cm处，也会堵塞输导组织，这与酶的活动有关。如果加入一些酸性物质（如柠檬酸）使水的pH值保持在3~4之间，可以抑制酶的活动，减轻这种生理堵塞，改善水分的供应环境，从而延长花期。

（4）抑制乙烯产生

花的衰老和内源激素变化的关系非常密切。乙烯和脱落酸具有促进鲜花衰老的作用，细胞分裂素则有延缓衰老的作用。其中，乙烯的合成是一个酶促反应过程，是一个需氧过程，高温会加速乙烯的产生。如果在切花保鲜液中加入一些抑制剂，如氨基氧乙酸（AOA）、氨基乙氧基乙烯基甘氨酸（AVG）、甲氧基乙烯基甘氨酸（MVG）等，可以抑制这一过程的进行。

important measures for the preservation of cut flowers is to store them in the environment with low temperature.

(2) Adding fungicide to preservative solution

Immersing cut flowers in water can not completely guarantee water supply, which is mainly due to their physiological causes and the blocked conducting system caused by microorganisms. Water often contains some microorganisms such as bacteria and fungi. They can invade plant vessels and secrete some harmful metabolites, and thus affect cut flowers' water absorption. So it is indispensable to add fungicide to preservative solutions so as to eliminate the negative impact of the micro-organisms.

(3) Adding acid materials to preservative solutions

Even if cut flowers are immersed in sterile water and free from the harmful impact of the microorganisms, they'll still exist poor water transportation. It is due to cut flowers' own physiological causes. The injured plant cells will produce secretion on the incisions of the bases of flowering shoots. For example, white milky liquid is produced on the incision of the sprays of *Euphorbia pulcherrima*. For some cut flowers, they will appear some sticky substances like oxides gathering around the incisions or 2 to 3cm above the surface of water and blocking the conducting tissues. The congestion is associated with the enzyme activities. If some acid substances like citric acid are added to water to keep the water pH value between 3 to 4, it can inhibit the enzyme activities, reduce the physiological congestion and improve the environment of water supply so as to extend the florescence.

(4) Inhibiting the production of ethylene

Flowers' aging is closely related to endogenous hormone changes. Ethylene and abscisic acid has the function of promoting the aging process while cytokinin has anti-aging effects. The synthesis of ethylene is a process of enzymatic reaction and also an aerobic process. High temperature can accelerate the production of ethylene. If some inhibitors, like amino oxyacetic acid (AOA), aminoethoxy vinyl glycine (AVG) and methoxyl vinyl

同时，银离子通过置换植物细胞内的乙烯，作用于受体中的铜离子，使乙烯不能参与作用，也会减弱乙烯的促衰老作用，如硫代硫酸银（STS），它可以极大地改善花材在花器中的保存期限，特别是对乙烯敏感的花朵，使用STS最恰当的时间是花材刚刚被剪切下来的时候。在使用中应注意，STS废液不要倒入下水道中，否则会污染环境。

（5）在保鲜液中加入糖

碳水化合物与蛋白质是鲜切花体内的一类重要物质。碳水化合物多以水溶性的单糖或双糖形式存在，有时以淀粉形式存在于细胞中。花期会消耗大量糖分，此时淀粉可转化为糖。糖是呼吸作用的重要原料，可提供维持切花生命活动需要的能量，也是合成多种有机酸的底物，具有保护细胞结构的功能。糖的含量和切花的品质有关，含糖量高的切花通常具有较高的观赏性与耐插性，所以在切花保鲜过程中，可以用外源糖补给切花，使之保持较长时间的寿命。

6.1.3 花材的保鲜方法

（1）扩大切口面积

将花枝剪切成斜切口，其斜度越大越好，或将切口纵向劈开，嵌入小石子撑开切口；也可以在靠近切口处，用刀切去部分皮层或用锤子将花茎末端3~4cm击碎，扩大吸水面积，加快吸水速率，以利于花枝吸收水分。此法对菊花、百合、马蹄莲、孔雀草等均适用。

glycine (MVG), are added to the preservative solutions, cut flowers' aging process can be inhibited.

Meanwhile, in the replacement of ethylene in plant cells, silver ions act on the copper ions of the receptors and make ethylene unable to function, which can reduce ethylene's role in promoting flowers' aging process. For example, the silver sulfate (STS) can greatly extend flowers' preservation in the vessels, especially for those flowers sensitive to ethylene. It is the most appropriate time to use STS when flowers are just cut down. STS wastes mustn't be poured into the sewers; otherwise, it may pollute the environment.

(5) Adding sugar to preservative solutions

Carbohydrates and proteins are very important substances in cut flowers. For the most part, carbohydrates exist in the form of water-soluble mono-saccharides or disaccharides, and sometimes in the form of starch in the plant cells. In the florescence, a large amount of sugar will be consumed when starch is converted into sugar. Sugar is a very important respiratory substrate to sustain the energy needed in the process of cut flowers' living activities. It is also the matrix to synthesize a variety of organic acids to protect the cellular structure. From the point of view of cut flowers, sugar content is related to the quality of cut flowers. Cut flowers with high sugar content are more ornamental and durable. As a result, in the process of the preservation of cut flowers, the exogenous sugar supply can be added to extend their life span.

6.1.3 Methods of Preserving Flower Materials

(1) Expanding the incision area

An incision is made into the flowering shoots in an oblique angle. The more oblique the angle is, the easier it is to split the incision vertically and embed small stones into it to open the incision. It is also proper to take such measures as to peel part of the spray near the incision with a knife or to smash with a hammer the tip of the stem, about 3 to 4 cm away from it. Beneficial to sprays absorbing water, these

（2）水中剪切法

把鲜切花茎端浸入水中，用剪刀在水中将其末端剪去1~2cm，以去掉导管内已存在的气泡，使切口在切离母体前不与空气接触，以免空气再侵入花枝基部导管，阻碍水分的吸收运输。注意一定要将花枝浸在水中，并且剪刀也伸到水下进行剪切，利用深水水压高及在水中导管不被空气堵塞的原理，可以使脱水花枝得以恢复。一般来说，花枝留取的高度可根据花卉吸水力的强弱确定，如唐菖蒲、姜等吸水力强，可少剪些；而吸水力弱的种类则应多剪些，如玫瑰、茉莉等。花枝剪切后，还应浸入桶中吸水15～20min后再拿出水面。此法简单易行，但不适合内含乳汁的鲜切花。

（3）烧

此法适合含乳汁和多肉的花材，如一品红等。这些花材剪切后切口处会流出汁液，将鲜切花茎端放到酒精灯上烧一下，再放入酒精中浸1min，最后放到清水中漂洗干净，这样可以阻止汁液外溢，避免切口、导管堵塞；否则，花朵会很快凋萎。同时，此法还可起到杀菌、防腐的作用。但烧时要注意不要损坏枝叶和花朵。月季、蜡梅、牡丹、芍药、丁香都可以采用这种方法。

measures can contribute to expanding the water absorption area, speeding up the water absorption and can be applied to *Dendranthema morifolium*, *Lilium* spp., *Zantedeschia aethiopic* and *Tagetes patula*.

(2) Underwater cutting method

Immersed into the water, 1 to 2 cm of the tip of the stem of cut flowers can be cut off with a knife to remove the bubbles in existence in the plant vessels. Before the incision is apart from the mother plant, make sure that it is not exposed to the air so that the air will not invade the plant vessels of the bases of the flowering shoots to hinder moisture absorption and transportation. In this process, sprays and a knife must be immersed into the water, and do it under the water with the knife. It is related to the principle of high water pressure in deep water and the principle that plant vessels under the water are not easily blocked by the air so that dehydrated sprays can recover. In general, the height of the sprays depends on the flowers' capacity of moisture absorption. For example, *Gladiolus gandavensis* and *Hedychium coronarium* (ginger lily) have great capacity of moisture absorption, their sprays can be cut less while those flowers like *Rosa rugosa* and *Jasminum* sambac have weaker capacity of moisture absorption, their sprays should be cut more. Having been cut off, the sprays should be immersed into a barrel for moisture absorption for 15 to 20 minutes. This method is simple, but it is not suitable for those cut flowers containing milk.

(3) Burning

This method is suitable for the flowers containing milk and pulp like *Euphorbia pulcherrima*. After these flowers are cut off, milk will come out of their incisions. The tips of the flower stems are to be burned on an alcohol lamp, immersed into alcohol for 1 minute, and then rinsed out in clean water. This method can prevent plant milk from spilling over to block the incisions and the vessels, otherwise, flowers are very likely to wilt soon. Meanwhile, this method also has the effect of sterilization and anti-corrosion. But take care not to damage

（4）烫

此法适用于草本花卉，尤其是吸水力差或含乳汁的草花，如晚香玉、菊花、非洲菊、香石竹、唐菖蒲、大丽花等，先将鲜切花上部的叶片及花保护好，以免水蒸气灼伤，然后将茎端 2～3cm 浸入 65～80℃ 的热水中浸泡 2～3min 或 100℃ 沸水中浸 10～30s，以浸烫部分刚好发白为宜，取出立即放入冷水中浸凉。此法可以排除导管内空气，有利于花枝吸水和杀菌。可用此法急救运输后因失水萎蔫的花材。

（5）切口药物处理法

在花枝切口涂抹酒精、盐、硼酸、乙酸、稀盐酸、煤油、薄荷油及一些洗涤剂，可以增强花枝的吸水性能，延长花期。

（6）远离催熟剂（乙烯）

过熟的果蔬及败落的花会释放大量乙烯气体。乙烯能加速花朵衰败，缩短花朵寿命。所以，鲜花摆放要远离苹果、香蕉、梨等水果；在花店中，要及时清理装有残枝败叶的垃圾桶，并保持通风良好。对于储存鲜花的保鲜柜，也要定期进行清理和通风换气。

（7）杀菌

在储存鲜切花的水中放入酒精、高锰酸钾、樟脑、硼酸、乙酸、柠檬酸、盐、明矾等物质都可以起到杀菌的作用。只有水质清洁，植物才能正常吸水保鲜。在水中放入杀菌物质时，一定要搅拌均匀，使其充分溶解，同时，注意浓度不可太高。

cut flowers' branches, leaves and blooms in the practice. This method is applicable to *Rosa chinensis*, *Chimonanthus praecox*, *Paeonia suffraticosa*, *P. lactiflora* and *Syringa* spp.

(4) Blanching

This method is suitable for herbaceous flowers, especially for those flowers with weaker capacity of moisture absorption or those containing milk, such as *Polianthes tuberosa*, *Dendranthema morifolium*, *Gerbera jamesonii*, *Dianthus caryophyllus*, *Gladiolus gandavensis* and *Dahlia pinnata*. First, leaves and blooms at the upper part of cut flowers should be well protected from being burned by steam. Second, 2 to 3cm of the tips of the stems should be immersed into hot water of 65 to 80℃ for 2 to 3 minutes, or soaked into boiling water of 100℃ for 10 to 30 seconds until they just turn white. Then remove and cool them immediately in the cold water. This method can get rid of the air in the plant vessels and is favorable for the moisture absorption and sterilization of the flowering shoots. As the first aid for those wilting flowers due to water loss in the process of transportation, it turns out quite practical.

(5) Chemical treatment on the incision

Alcohol, salt, boric acid, acetic acid, dilute hydrochloric acid, kerosene, peppermint oil and some detergents can be applied to the incisions of the flowering shoots to enhance their capacity of moisture absorption and thus to extend their florescence.

(6) Keeping away from the ripening agent (ethylene)

Overripe fruits and vegetables and the wilted flowers will release a lot of ethylene gases, which can accelerate the process of flower wilting and shorten their life span. Therefore, flowers should be far away from such fruits as apples, bananas and pears. In flower stores, trash cans containing the wilted branches and leaves should be emptied in time and the stores should remain in the well-ventilated condition. It is also true of the cupboards to preserve flowers and it is advisable to clean and ventilate them regularly.

(7) Sterilization

In the water to preserve cut flowers, such substances

（8）补充营养

鲜花从母体上切离以后，就失去了营养来源，在储存鲜花的液体中加入糖、啤酒、阿司匹林、维生素C等各类营养物质，可以提供一定的营养，使鲜切花保持花色鲜艳，延长花期。

（9）调节储存温度

原产于热带的鲜花，如花烛、鹤望兰等，需要保存在10~15℃的环境中。而原产于温带的一些花卉，如切花月季、菊花等，应保存在5℃左右的环境中。温度过高，鲜切花代谢作用较强，容易凋谢，所以应尽快将鲜花冷藏。

6.1.4 花卉保鲜剂

（1）花卉保鲜剂的含义

花卉保鲜剂是指用于调节切花生理生化过程，抵抗外界不良环境变化，保存切花的良好品质，延迟衰老的化学药剂。目前主要是制成溶液使用。花卉保鲜剂包括一般保鲜液、水合液、脉冲液、STS脉冲液、花蕾开放液和瓶插保持液等。在采后处理的各个环节，从栽培者、批发商、零售商到消费者，都可以使用花卉保鲜剂。花卉保鲜剂能使花朵增大，保持叶片和花瓣的色泽，从而提高花卉品质，延长货架寿命和瓶插寿命。

as alcohol, potassium permanganate, camphor, boric acid, acetic acid, citric acid, salt and alum can be added for the effect of sterilization. Only if clean water is guaranteed, can plants normally absorb water for their own preservation. When these substances are added to the water for sterilization, they must be mixed properly and dissolved adequately. Furthermore, moderate concentration is advisable.

(8) Nutritional supplements

After being cut off from their own mother plants, cut flowers lose their sources of nutrition. Such nutrients as sugar, beer, aspirin and vitamin C can be added to the liquid in which cut flowers are preserved. In this way, a certain amount of nutrition can be supplied to keep flowers in their bright colors and extend their florescence.

(9) Storage temperature adjustment

Such tropical flowers as *Anthurium* spp. and *strelitzia reginae* need to be preserved in the environment of about 10 to 15℃ while temperate flowers like *Rosa chinensis* and *Dendranthema morifolium* need to be preserved in the environment of about 5℃. In too high temperature, cut flowers will easily wilt due to relatively stronger metabolism. As a result, cut flowers should be stored in lower temperature as soon as possible.

6.1.4 Preservative Solutions for Flowers

(1) Definition of preservative solutions for flowers

Preservation solutions for flowers are the chemicals to adjust cut flowers' physiological and biochemical processes, to resist the external unfavorable environmental changes, to ensure flowers' quality and to delay their process of wilting. At present, they are mainly made in the form of solutions, including the common preservative solutions, hydration solutions, pulse solutions, STS pulse solutions, blooming solutions and vase flower preservative solutions. At every stage of processing, preservative solutions for flowers are quite useful to growers, wholesalers, retailers and consumers. They can enlarge flowers, keep the bright colors

（2）花卉保鲜剂的主要成分和作用

大部分商业性保鲜剂都含有碳水化合物、杀菌剂、抑菌剂、乙烯抑制剂、生长调节剂和矿质营养成分等，以下分别详述它们的作用。

①碳水化合物：保鲜剂中使用的碳水化合物主要是蔗糖、葡萄糖和果糖。外供糖源作为切花的主要营养源和能量来源将参与延长瓶插寿命的基础过程，保持细胞中线粒体结构并维持其功能，调节蒸腾作用和细胞渗透压，促进水分平衡，增强水分吸收，延迟蛋白质的水解。

不同的切花种类，或同一种类不同品种所用保鲜液中糖的浓度不同。如在花蕾开放液中，香石竹最合适的糖浓度为10%，而菊花叶片对糖浓度敏感，一般用2%的浓度。对于月季切花，高于1.5%的糖浓度易引起叶片烧伤。叶片对高浓度的糖比花瓣反应更敏感，可能是因为叶细胞渗透压调节能力差的缘故。因此，提高糖浓度的限制因子往往是叶片的敏感性。在实践中，大部分保鲜剂使用相对低的糖浓度，以避免造成伤害，但碳水化合物量不足就不能最大限度地提高品质，延长瓶插寿命。因此，糖浓度是关键因子。一般来讲，对特定切花，保鲜剂处理时间越长，所需糖浓度越低。

保鲜剂中的糖也是微生物生长的最佳基质，微生物繁殖过多又引起花茎导管的阻塞。因此，在保鲜剂中糖与杀菌剂应结合使用。

of both their leaves and petals, and thus improve cut flowers' quality and prolong their shelf life and vase life.

(2) Main constituents and functions of preservative solutions for flowers

Most commercial preservative solutions contain carbohydrates, fungicides and antibacterial agents, ethylene inhibitors, growth regulators and mineral nutrients. Their functions will be respectively dealt with in the following:

① Carbohydrates: Carbohydrates used in preservative solutions for flowers are mainly sucrose, glucose and fructose. Serving as the main sources of cut flowers' nutrition and energy, these external sources of sugar will participate in the fundamental process of prolonging cut flowers' vase life span. They can keep the cell mitochondria structure, maintain its function, adjust the transpiration and cell osmotic pressure, improve moisture balance, increase water absorption, and delay the hydrolysis of proteins.

According to the various genera of cut flowers or the various breeds of the same genus, the sugar concentration in the preservative solutions for cut flowers also varies accordingly. For example, in blooming solutions, the most appropriate sugar concentration is 10% for *Dianthus caryophyllus*; whereas, the appropriate concentration is generally 2%, for *Dendranthema morifolium* leaves are sensitive to sugar concentration. For *Rosa chinensis*, the leaves are easily burned if the sugar concentration is above 1.5%. Leaves are more sensitive to high sugar concentration than petals. It is possibly due to leaf cells' less regulatory capacity to osmosis. Therefore, the inhibitory factor to improve sugar concentration tends to be leaves' sensitivity. In practice, most preservative solutions contain comparatively lower sugar concentration to avoid damage to cut flowers. But if there are not sufficient carbohydrates, the maximum quality improvement and the prolonged life span can't be achieved. Hence the key factor is sugar concentration. On average, for a particular cut flower, the longer the preservation solution is applied, the lower the required sugar concentration is.

②杀菌剂和抑菌剂：在花瓶水中生长的微生物种类有细菌、酵母和霉菌。这些微生物大量繁殖后阻塞花茎导管，影响切花吸收水分，并产生乙烯和其他有毒物质而加速切花衰老，缩短切花寿命。例如，当水溶液中细菌浓度达到 $10^7 \sim 10^8$ 个/mL，就引起月季花茎吸水力下降，当细菌浓度达到 3×10^9 个/mL 时，在 1h 内切花开始出现萎蔫。此外，在植物组织中，细菌还可增强切花在储藏期间对低温的敏感性。

为了抑制微生物生长，保鲜剂中可加入杀菌剂或与其他成分混用。最常用的杀菌剂是 8-羟基喹啉盐类。它们是广谱杀菌剂和杀真菌剂。8-羟基喹啉硫酸盐（8-HQS）和 8-羟基喹啉柠檬酸盐（8-HQC）可使保鲜液酸化，有利于花茎吸水，延长瓶插寿命。但是 8-羟基喹啉在一些切花中会引起负作用，如它可以造成菊花和丝石竹的叶片烧伤和花茎褐化，因此限制了其广泛应用。其他杀菌剂还有硝酸银、醋酸银、硫代硫酸银、硫酸铝、缓释氯化合物、乙酸、食盐、山梨酸、苯甲酸等。

③乙烯抑制剂：硫代硫酸银（STS）是目前花卉业使用最广泛的乙烯抑制剂。STS 的生理毒性较硝酸银低，在植物体内有较好移动性，易于从花茎移至花冠，对花朵内乙烯合成有高效抑制作用，并使切花对外源乙烯作用不敏感，可有效地延长多种切花的瓶插寿命。它不易被固定，在较低浓度时就起作用。用 1~4mmol 的 STS 处理香石竹、百合和其他切花 5min 至 24h，就可明显地抑制它们的衰老过程。可用

Sugar in preservative solutions is also the best stroma. Seeing that too much microbial reproduction will cause the obstruction of plant stem vessels, sugar and fungicides should be combined in preservative solutions.

② Fungicides and antibacterial agents: There grow a variety of microorganisms in the water of vases such as bacteria, yeasts and molds. Their reproduction will jam stem vessels, affect moisture absorption of cut flowers, and produce ethylene and other toxic substances so as to accelerate their aging and shorten their life span. For example, when the bacterium density in the water of vases arrives at 10^7 to 10^8 cfu per milliliter, it will decrease Rosa chinensis capacity of moisture absorption; when the bacterium density reaches 3×10^9 cfu per milliliter, cut flowers will begin to wilt within an hour. And what's more, in plant tissues, bacteria can increase cut flowers' sensitivity to low temperature during their storage.

To inhibit the growth of microorganisms, fungicides can be added to preservative solutions or mixed with other ingredients. The most commonly used fungicides are salt-containing 8-hydroxyquinolines. They belong to broad-spectrum fungicides and fungicides. 8-hydroxyquinoline sulfate (8-HQS) and 8-hydroxyquinoline citrate (8-HQC) can cause the acidification of preservative solutions, which is beneficial to the moisture absorption of stems and can prolong vase life span of cut flowers. However, 8-hydroxyquinoline may exert negative effects on some cut flowers. For example, it may lead to leave burns and stem browning of *Dendranthema morifolium* and *Gypsophila paniculata* and hence the limitation of its widespread application. The other fungicides include silver nitrate, silver acetate, silver thiosulphate, aluminum sulfate, slow-released chlorine compounds, acetic acid, salt, sorbic acid and benzoic acid.

③ Ethylene inhibitors: Silver thiosulphate (STS) is the best widely used ethylene inhibitor in current flower business. It has lower physiological toxicity than silver nitrate. Its good mobility in plants facilitates it to move

STS阻止金鱼草、翠雀和香豌豆小花的乙烯诱导脱落。但STS浓度过高或处理时间过长会对花瓣和叶片造成损害。

STS需随配随用，配制方法如下：先溶解0.079g硝酸银（AgNO₃）于500mL无离子水中，再溶解0.462g硫代硫酸钠（Na₂S₂O₃·5H₂O）于500mL无离子水中，把AgNO₃溶液倒入Na₂S₂O₃·5H₂O溶液中，并不断搅拌。此混合液即为银离子浓度为0.463mmol的STS溶液。配好的溶液最好立即使用，如不马上使用，应避光保存在棕色玻璃瓶或暗色塑料容器内。STS溶液可在20~30℃黑暗环境中保存4d。

④生长调节剂：生长调节剂也用于花卉保鲜剂中，它们包括人工合成的生长激素和阻止内源激素作用的一些化合物。植物生长调节剂可单独使用或与其他成分混用。生长调节剂可引起、加速或抑制植物体内各种生理和生化进程，从而延缓切花的衰老过程。

细胞分裂素是最常见的保鲜剂成分。细胞分裂素的主要作用在于降低切花对乙烯的敏感性，抑制乙烯产生，从而延长切花寿命。它可抑制紫罗兰、微型唐菖蒲叶片的黄化。

from its stems to its corollas, inhibits effectively ethylene synthesis and makes cut flowers less sensitive to exogenous ethylene actions so as to effectively prolong the vase life span of most cut flowers. It is not easily fixed and works in comparatively lower concentration. After *Dianthus caryophyllus*, *Lilium* spp. and other cut flowers are treated with 1 to 4 micromoles of STS for 5 minutes to 24 hours, their aging can be obviously inhibited. STS can also prevent the ethylene induction losses of the florets of *Antirrhinum majus*, *Delphinium grandiflorum* (delphinium) and *Lathyrus odoratus*. But too high concentration of STS or too long treatment may cause damage to cut flowers' petals and leaves.

STS needs to be used immediately after being made up. Its preparation method is as follows: firstly, 0.079 grams of silver nitrate ($AgNO_3$) should be dissolved in 500 ml of deionized water, and then 0.462 grams of sodium thiosulphate ($Na_2S_2O_3·5H_2O$). Secondly, pour the $AgNO_3$ solution into the $Na_2S_2O_3·5H_2O$ solution and stir them constantly. The resulted mixture is the STS solution with 0.463 millimoles of silver ion concentration. It is advised to use the prepared solution immediately or to preserve it in brown glasses or dark plastic containers without being exposed to light. In the dark environment of 20 to 30 ℃, STS solutions can be preserved for 4 days.

④ Growth regulators: Growth regulators are also used in preservative solutions for flowers, including synthetic growth hormones and some compounds preventing the endogenous hormone effects. Used alone or mixed with other ingredients, growth regulators can cause, speed up or inhibit a variety of physiological and biochemical processes in plants, thus delaying their aging process.

Cytokinin is the most commonly used ingredient in preservative solutions. Its main function lies in its capacity to reduce cut flowers' sensitivity to ethylene so that their life span can be prolonged. For example, it can inhibit the yellowing of the blades of *Matthiola incana*(violets) and miniature *Gladiolus gandavensis*.

利用细胞分裂素处理切花可采用喷布或浸蘸的方法，其有效浓度因处理时间而异，0.001%~0.01%可用于瓶插保持液和花蕾开放液的长时间处理，0.01%可用于较短时间的脉冲处理，0.025%用于整个花茎浸蘸处理2min。浓度过高，处理时间过长，也会产生不良后果。

细胞分裂素（BA）处理香石竹效果最好，它不仅能使切花对乙烯的敏感性降低，同时还使乙烯释放高峰推迟，从而延长采后寿命。对月季、鸢尾和郁金香的处理效果也不错。用0.001%BA短时浸蘸花烛，可延长其保质期，增强对冷害的抵抗力。非洲菊切花在0.0025% BA中只需浸泡2min就可阻碍切花水分丢失，延迟细胞中离子渗漏，延长采后寿命。水仙切花在0.01% BA和0.0022% 2,4-D混合液中浸沾可延迟衰老。0.0005% BA和0.002% NAA混合液处理可加快储后香石竹花蕾的开放。

常用的生长延缓剂有比久（B$_9$）和矮壮素（CCC），它们可延长切花采后寿命。这些化合物抑制植物生长，阻止组织中赤霉酸生物合成及其他代谢过程，因此增加了切花抗逆性。它们的有效浓度因植物种类和季节而异。B$_9$的适宜浓度为金鱼草0.001%~0.005%、紫罗兰0.0025%，香石竹和月季0.005%。含0.005%CCC的花瓶保持液（内还含有8-HQS和蔗糖）可延长郁金香、香豌豆、紫罗兰、金鱼草和香石竹的瓶插寿命。

Cut flowers can be treated with cytokinins by way of spraying or dipping. Its effective concentration varies with different amounts of treatment time. Cytokinin with the effective concentration between 0.001% and 0.01% is used to process flowers with vase flower preservative solutions and blooming solutions for a long time. Cytokinin with that of 0.01% can be used in pulse treatment within a short period of time. And cytokinin with that of 0.025% is used for the 2-minute treatment of the stems by way of dipping. In addition, too high concentration or too long treatment will result in negative consequences.

Cytokinins works best when used to process *Dianthus caryophyllus*. It can not only reduce cut flowers' sensitivity to ethylene, but also delay ethylene's peak of releasing and thus prolong their life span. To *Rosa chinensis*, *Iris tectorum*(iris) and *Tulipa gesneriana*, it also works well. Dipping *Anthurium* spp. into 0.001% BA of cytokinins for a short period of time can prolong their shelf life span and increase their capacity to resist chilling injuries. Dipping *Gerbera jamesonii* in 0.0025% BA just for 2 minutes can prevent their water loss, delay their cell ion leakage and prolong their shelf life span. Dipping into 0.01% BA of and 0.0022% ppm 2,4-D of mixtures can delay the aging of *Narcissus tazetta* var *chinensis*. 0.0005% BA of and 0.002% NAA of mixtures can accelerate *Dianthus caryophyllus* blooming.

Commonly-used growth retardants include alar (B$_9$) and cycocel (CCC), which can prolong cut flowers' shelf life span. These compounds inhibit plants' growth, and prevent their biosynthesis of gibberellic acid and other metabolic processes. Therefore, they enhance cut flowers' endurance of the adversities. Their effective concentration varies with their breeds and seasons. The effective concentration of growth retardant B$_9$ is 0.001% to 0.005% (*Antirrhinum majus*), 0.0025% (*Matthiola incana*) and 0.05% (*Dianthus caryophyllus* and *Rosa chinensis*). Containing 8-HQS and sucrose, vase preservative solutions with the 0.005% concentration of CCC can prolong the vase life span of *Tulipa gesneriana*, *Lathyrus odoratus*, *Matthiola inoana* and

⑤有机酸：有机酸能降低保鲜剂的 pH，促进花茎水分吸收和平衡，减少花茎的阻塞。用于保鲜剂的有机酸有柠檬酸、异抗坏血酸、酒石酸和苯甲酸，其中应用最广的是柠檬酸。

柠檬酸的使用浓度为 0.005%～0.08%，它对改善月季、菊花、羽扇豆、唐菖蒲、鹤望兰的水分吸收效果尤佳。苯甲酸（浓度 0.05%）可有效延长花烛的寿命。

0.015%～0.03% 苯甲酸钠可延迟香石竹和水仙的衰老，但对金鱼草、鸢尾、菊花和月季没有作用。苯甲酸钠作为抗氧化剂和自由基清除剂，可减少切花乙烯产生，并增加水溶液的酸度，有利于延长一些切花的采后寿命。

异抗坏血酸或抗坏血酸钠有效浓度为 0.01%，具有抗氧化功能和生长促进作用。可作为花瓶保持液阻止月季、香石竹和金鱼草的衰老。

在实际生产中，花卉保鲜剂的主要成分是糖和杀菌剂，有时再增加 1～2 种成分。一种典型的保鲜液可能含有 1% 的蔗糖，一种杀菌剂（0.02% 8-HQS，或 8-HQC 或 0.005% 硝酸银），一种酸化剂（0.02%～0.06% 柠檬酸或硫酸铝）。

Dianthus caryophyllus.

⑤ Organic acids: Organic acids can reduce the pH value of preservative solutions, improve stem moisture absorption and their balance, and reduce the obstruction to their stems, what can be used in preservative solutions include citric acid, ascorbic acid, tartaric acid and benzoic acid, of which the most widely-used is citric acid.

The effective concentration of citric acids is 0.005% to 0.08% and they work best in improving the moisture absorption of *Rosa chinensis*, *Dendranthema morifolium*, *Lupinus micranthus*, *Gladiolus gandavensis* and *Strelitzia reginae*. Benzoic acid with the concentration of 0.05% can effectively prolong the life span of *Anthurium* spp.

Sodium benzoates with the concentration of 0.015% to 0.03% can delay the aging of *Dianthus caryophyllus* and *Narcissus tazetta* var. *chinensis*, but do not work well to *Antirrhinum majus*, *Iris tectorum*, *Dendranthema morifolium* and *Rosa chinensis*. As an anti-oxidant and a free radical remover, sodium benzoate can reduce the production of ethylene in cut flowers and increase the acidity of the solutions, which is quite favorable to some cut flowers' shelf life span.

With the effective concentration of 0.01%, ascorbic acid and sodium ascorbate have the function of anti-oxidation and the effect of promoting plant growth. As a result, they can be used as vase preservative solutions to prevent the aging process of *Rosa chinensis*, *Dianthus caryophyllus* and *Antirrhinum majus*.

In production, the main ingredients of flower preservative solutions are sugar and fungicides, with 1 to 2 extra ingredients sometimes. A typical preservative solution may contain 1% of sucrose, a fungicide (8-HQS or 8-HQC with the concentration of 0.02%, or silver nitrate with the concentration of 0.005%), and a kind of acidifier (citric acid or aluminum sulfate with the concentration of 0.02%～0.06%).

6.2 花卉包装

花卉包装是指根据鲜花特性，采用适宜的包装材料或花器，将花卉商品包封或盛装，以达到保护花卉商品、方便储运、促进销售的目的。鲜花经过包装后，可以安全完好地到达顾客手中；同时，包装作为一种广告营销工具，可以进行信息传递，有利于顾客识别商品，激发购买欲望，也是花店营销中一项重要的内容和手段。良好的包装，必须定位准确，符合消费者审美观，能够促使消费者在琳琅满目的商品中迅速地进行甄选和识别，从而对产品品牌形成记忆，促成消费者的购买行为。如果不重视花卉的包装，会降低花卉商品的市场竞争力。包装设计由标志、颜色、形状、规格等要素组成，在进行花卉商品的包装设计时应定位准确，具有独特性，与其他花卉商品有所区别。

花卉包装是花卉商品的有机组成部分，在销售的过程中是随花卉商品一同出售给顾客的，好的包装可以保持花卉商品完好无损，防止插花中水分的渗漏，保持花卉商品整洁，便于顾客选购和携带。

6.2.1 花卉包装与顾客心理

（1）花卉包装的功能

花卉商品的包装最开始是用来承载和保护花卉商品，以避免其损坏、散落及防止水分溢出。包装对消费者心理有巨大的影响，甚至可以左右他们对花卉商品的认识和感受，可以称为"沉默而极有说服力的推销员"。包装的功能有以下几点：

6.2 Flower Packaging

Flower packing refers to how to adopt suitable packing materials or vessels to pack or contain flower commodities according to their characteristics, for the purpose of protecting flowers, making their storage and transportation convenient and expanding their sales. In the process of their storage and transportation, packaged flowers can reach customers safe and sound; at the same time, as a commercial marketing tool, packaging can convey information, facilitate customers to identify goods, stimulate consumption desires and is also an important component and means of the marketing strategies of flower shops. Good packaging must be accurately strategy-oriented, able to conform to consumers' aesthetic views and promote consumers to select and identify rapidly in a wide variety of goods available. Thus a certain brand will be rooted in consumers' mind, which can lead to consumers' consumption. If flower packaging is not treated seriously, it will reduce the marketing competitiveness of flowers. Packaging design consists of such key elements as logo, color, shape and specification. It should be accurately orientated, unique and make some flower product distinguishable from the others.

Flower packaging is an integral part of flower commodities, sold to customers along with flower commodities. Good packaging can keep flower commodities safe and sound, prevent the water leakage of cut flowers, maintain them tidy and make it convenient for customers to select and carry.

6.2.1 Flower Packaging and Consumer Psychology

(1) Functions of Flower Packaging

At the very beginning, flower commodity packaging was used to carry and protect flowers in order to avoid their damage, scattering and water overflow. Referred to as "a silent and most convincing salesman", it has a great impact on consumer psychology and can even determine consumers'

①识别功能：在花卉市场上，质量接近的花卉种类很多，采用不同的包装可以使它们产生差异。设计精良、富于审美、独具特色的包装能使花卉商品在众多的同类商品中脱颖而出，以其独特的魅力吸引消费者的注意力并给消费者留下深刻印象。同时，花卉批发商在包装上印刷准确而详尽的文字说明，能够全面显示花卉商品的产地、特色、联系方式等重要信息，有利于今后扩展业务。

②便利功能：好的花卉包装不仅使人耳目一新，还可以有效地起到保护作用，有利于花卉商品的储存，延长花卉商品的寿命。花卉商品经过包装，还有利于消费者选购和携带。因此，需要根据实际，设计合理、便利的花卉商品包装。

③美化功能：设计具有艺术性的花卉商品包装，为花卉商品披上美丽的色彩，这样能更有效地刺激消费者的感官，使花卉商品锦上添花、赏心悦目，从而有效地推动顾客购买，并且能够提高花卉商品的观赏价值，提高花卉商品的档次和竞争力，起到美化、宣传作用。

④增值功能：消费者选购商品时对商品价值的感受往往是从包装开始。设计成功的花卉商品包装，会融艺术性、知识性、趣味性和时代性于一身,能满足消费者的自我表现心理，让消费者在拥有花卉商品的同时在精神上得到极大满足。

understanding and feelings of flower commodities. The functions of flower packaging are as follows:

① Identification: In the flower market, there are a wide variety of flowers with similar quality and different packaging can make great difference to them. The aesthetic and unique packaging with an extraordinary design can make a particular flower product stand out among many other similar products, attract consumers' attention with its unique charm and leave a deep impression on them. Meanwhile, flower wholesalers print accurate and detailed texts on the packaging, which can fully display such important information as flower products' origin, characteristics and contact information so as to bring more business in the future.

② Convenience: Good packaging can not only make people find everything new and fresh, but also be effective in environmental protection and favorable to storing flowers and extending their life span. Through packaging, it is also convenient for consumers to select, purchase and carry flowers. Therefore, it is necessary to design reasonable and convenient flower packaging according to pragmatic purposes.

③ Beautification: Artistic flower packaging can add beautiful colors to flower products and more effectively stimulate consumers' sensory organs, which makes flowers more perfect and pleasing to both the eyes and the mind of the consumers. Thus it can effectively promote customer consumption, increase flowers' ornamental values, goods grade and competitiveness, and play a part in their beautification and publicity.

④ Value-added function: When consumers choose and purchase goods, their feelings for goods value often begin with flower packaging. Successfully-designed flower packaging will integrate artistic quality, knowledge, interests and times, satisfy consumers' self-expression psychology and enjoy great spiritual satisfaction in possession of flower commodities.

（2）花卉包装对顾客心理的作用过程

①唤起注意：花卉商品包装的首要功能是引起顾客的注意，吸引消费者的眼球。市场经济被称为眼球经济，符合大众主流审美观的产品包装，能在外观上对消费者形成一种强有力的视觉冲击，吸引消费者的关注。作为消费者刺激的重要表现形式，不同包装物给予顾客的刺激强度有明显差异。

②引起兴趣：花卉商品包装要与消费者的偏好相联系，要引起消费者对花卉商品的兴趣。因此，花店在进行花卉商品包装设计时需要考虑消费者的年龄、性格、职业、文化、经济状况。只有充分研究顾客的兴趣偏好，使包装与花卉商品的风格一致并符合顾客的价值标准，才能打动消费者，引起消费者对花卉产品的兴趣。

③启发欲望：启发欲望就是刺激顾客需求。顾客在产生购买动机后，其购买行动的最终实现还要取决于对刺激物的感受。对于琳琅满目的花卉商品，在质量和价格差异较小时，消费者会直接根据自身的喜好和产品的外在包装，在心理上对产品进行主观的等级分类。因此，包装外观如果能在感官上与消费者的审美、取向、偏好等方面形成共鸣，就能使花卉产品在潜在顾客中形成好感。

④导致购买：这是花卉商品包装对顾客心理作用的最终目的。因此，花卉包装应具有推销功能，别具一格，加上包装上的大量信息，吸引顾客，使顾客爱不释手，最终决定购买。

(2) Effects of Flower Packaging on Consumer Psychology

① Drawing attention: The primary function of flower product packaging is to attract customers' attention. This kind of market economy is known as eye-catching economy. Product packaging fitting in with the mainstream aesthetic views can exert a powerful visual impact on consumers and attract their attention. As an important form of expression to stimulate consumers, different packaging has obvious difference in stimulation strength to customers.

② Arousing interest: Flower product packaging is supposed to be linked with consumers' preferences, to attract their interests in flower commodities. Therefore, in designing flower product packaging, flower shops should consider consumers' age, character, career, cultural background and economic conditions. Only by fully researching customers' interests and preferences, can packaging fit in with flower commodities in their styles and conform to customers' value standards so that it can impress consumers and attract their interests in flower commodities.

③ Inspiring desire: Inspiring desire is to stimulate consumer demand. After purchasing motivation is stimulated, customers' actual final purchasing depends on their feeling for stimuli. In their choice of a wide variety of flower commodities, when there are comparatively slight differences in product quality and prices, consumers will psychologically and subjectively sort out product grades according to their own preferences and product exterior packaging. As a result, if the product's exterior packaging can resonate with customers in such aspects as their aesthetic views, orientations and preferences, it will enable flower commodities to positively impress potential customers.

④ Making purchases: Making purchase is the ultimate goal of the psychological effects of flower product packaging on customers. Flower product packaging should have marketing functions. A unique packaging style with a lot of information on the packaging tends to attract customers' eyes, make them fondle the flower product admiringly, and

（3）花卉包装的要求

因为不同的国家和地区有不同的风俗习惯和价值观念，有他们自己喜爱和禁忌的人物、动物、植物、图案，在进行花卉商品包装设计时只有考虑这些，才有可能赢得当地市场的认可。花卉商品的包装设计应符合以下心理要求：

①色彩搭配协调：色彩作为顾客接触花卉商品时第一关注的要素，在花卉包装设计中占有特别重要的地位。因为花卉本身具有特定的颜色，在选择包装材料色彩时既需要搭配协调，又要突出该种花卉商品特有的视觉特征，使之更富有诱惑消费者的魅力，刺激和引导消费，同时增强人们对品牌的记忆。包装色彩要有清楚的识别性及较高的明视度，要与其他设计因素和谐统一，以便被花卉商品的购买人群所接受。色彩在不同市场、不同陈列环境都应充满活力，同时，要注意单个包装的效果与多个包装的叠放效果。

②符合花卉商品的特征：鲜花作为脱离母体的活的花枝，仍在进行着有限的生命活动，所以在进行包装时要注意选择防水材料，并考虑到鲜花的保鲜。同时，要采用适当的包装形式和包装材料使鲜花的包装形象突出，别具一格。

③方便消费者：花卉商品经过包装后，更有利于消费者的观察、挑选、购买、携带和使用。如果采用透明、半透明的包装，消费者能对花卉商品获得更直观、鲜明、真实的体验。

thus lead them to make purchases.

(3) Demands in the Flower Packaging

In the design of flower product packaging, it should be noted that different countries and regions have different customs and values, and that they are fond of some particular figures, animals, plants and patterns and have others under taboo. Only by adapting to these requirements can flower product packaging win the approval of local markets. The design of flower product packaging should meet the following psychological requirements:

① Harmonious color arrangement: When customers begin to contact flower products, it is color that first comes into their view and color occupies an especially important position in the design of flower product packaging. Since each species of flowers has its own particular color, it is necessary to coordinate different colors and highlight the visual characteristics of a particular product. It will be more appealing to consumers, stimulates and guides consumption, and leaves a deep impression of product brands on people. Packaging color should have clear recognition and high visibility, and be in harmony with the other design factors, in order to be accepted by the purchasers. It should be dynamic in various markets and display environments. And what's more, the effect of individual packaging and of stacked multiple packaging deserves equal attention.

② Conforming to the characteristics of flower products: As the living flowering shoots off the mother plant, cut flowers still carry on their limited living activities, so it should be careful to choose water-proof materials in packing flowers and flower preservation should also be considered. In the meantime, it is advisable to adopt the appropriate forms of packaging and packaging materials to make the image of flower packaging stand out and unique.

③ Offering convenience to consumers: After flower products are packaged, it is more advantageous for consumers to observe, select, purchase, carry and use. If the packaging is transparentor or translucent, consumers can achieve a more intuitive, vivid and real psychological

④具有时代气息：随着时代发展，人们的审美观、价值取向都会发生一些转变。在进行花卉包装时要顺应最新潮流，对包装材料的选用、包装技巧、样式、色彩搭配等方面进行调整，提供给顾客新颖独特、简洁明快的花卉包装。

⑤具有针对性：不同的消费者由于教育背景、收入水平、生活方式、消费习惯及购买目的不同，对花卉商品包装的要求也不同。有的追求廉价实用，有的喜爱美观大方，有的则要求豪华高档。因此，在花卉商品的包装设计中应强调针对不同的消费群体采用不同的包装风格。

6.2.2 现代包装材料

随着经济的发展和人们生活质量的提高，现阶段用于花卉商品的包装手段和方法也呈现多样化的趋势。实际操作中，要根据不同的花卉商品、档次、用途及消费对象、风俗、销售范围、方式等采用不同的包装材料，并进行包装设计。花卉包装在经济、牢固、美观的基础上还担任着"无声的售货员"，因此设计主题鲜明、风格独特、具有吸引力的包装有利于花卉商品的销售。插花员要根据具体的花卉商品采用相应的包装方法和材料，并贴上花店的商标，树立花店的品牌。

伴随鲜花市场的不断繁荣，鲜花包装市场也日益红火。包装方式多样化，有塑料包装、纸包装、纸盒包装、尼龙纱网包装等，既有国产的也有进口的包装材料。这些异彩纷呈的包装材料和包装手段，把鲜花衬托得更加妩媚动人。

④ Reflecting the features of modern times: People's aesthetic views and value orientation change with the development of the times. Therefore, flower product packaging should fit in with the latest trend of the times and make according adjustments in the choices of packaging materials, packaging skills, packaging styles, color matching and so on, to provide customers with novel, unique, concise and lively packaging.

⑤ Being oriented: Due to the various education backgrounds, income levels, lifestyles, consumption habits and purchasing purposes, customers hold different requirements for flower product packaging. Some prefer cheap and practical packaging, some favor packaging elegant and in good taste, while others desire luxurious and high-graded packaging. Therefore, it is emphasized that various packaging styles should be adopted in packaging design, targeted at various groups of consumers.

6.2.2 Modern Packaging Materials

With the development of economy and the improvement of people's life quality, there is a trend of diversified means and methods of flower product packaging at present. In practice, packaging materials and packaging designs should vary with different flower products, grades, purposes, groups of consumers, customs, sales scopes, sales methods and the like. Flower product packaging plays a role of "a silent salesman" on the basis of being economical, secure and elegant. So the packaging with a distinct, unique and attractive design theme is beneficial to flower sales. Flower-arranging staff should adopt various packaging methods and materials according to flower products, and attach the trademarks of flower shops to set up flower shop brands.

With the increasing prosperity of flower markets, flower packaging markets are prosperous likewise. There are a wide variety of packaging methods including plastic packaging, paper packaging, carton packaging and nylon gauze packaging, made in China and abroad. Flowers

（1）礼盒包装

盒子的包装方法大致分为两种：一种是把扎好的花束进行外包装之后，放到精美的盒子里，这种包装方法常用于长方形的盒子；另一种是直接根据盒子的形状，把花摆放在盒子里，可以摆成心形或其他形状，甚至文字，常用于正方形和心形的盒子。

礼盒有很多种材料，如纸盒、塑料盒等。花店中常用纸盒包装，如单枝玫瑰盒、12枝玫瑰盒。

①纸盒包装鲜花的优点：

重量轻，携带方便。

洁净、环保、可防尘。

纸盒便于机械化生产，易实现包装标准化。

用盒子包装鲜花，形式新颖、美观。

易于折叠，存放时节省空间。

另外，若在盒子底部垫上塑料纸，放好花泥后，还可以将花头向上插置，这样可以充分利用不同长度的花枝，如花店在采购或整理鲜花时碰断的短花枝，可以在这种设计中充分利用，以节约成本。需要注意的是，此种花艺设计形式一定要做好防水，以免纸盒被水浸湿。目前有一种塑胶盒，常用于这种花艺设计。

②纸盒包装鲜花的缺点：

把鲜花放在盒子里，花枝不透气、不见光，不利于鲜花保鲜。

纸盒的硬度较低，怕挤压。

纸盒怕水浸，在盛放或插置鲜花时，需做好防水处理。

stand out more charming against these colorful packaging materials and packaging methods.

(1) Box Packaging

Box packaging methods are roughly divided into two different kinds: one is to put bouquets into delicate boxes after bundling and packaging them, which is often applicable to rectangular boxes; the other is to put directly into boxes bouquets in the shape of a heart or something, or even words according to the boxes' shapes, which is often applicable to square and hear-shaped boxes.

There are a lot of materials for box packaging, such as cartons and plastic boxes. The relatively commonly-used materials in flower shops are cartons, for example, single-rose cartons and 12-rose cartons.

① Advantages of packaging flowers with cartons:

Box packaging is light and convenient to carry.

Box packaging is clean, environmentally-friendly and dust-proof.

Box packaging is convenient for mechanized production and the implementation of packaging standardization.

Packaging flowers with boxes is novel and elegant.

Boxes are easy to fold and to save space in storage.

If the bottom of packaging boxes is padded with plastic paper, flowers can be transplanted after floral clay is placed. In this way, flowering shoots with different length can be made full use of to save costs if short flowering shoots are broken off in the process of purchasing or arranging cut flowers. It is important to note that this kind of flower art design should take water-proof matters into consideration in order to prevent boxes from getting wet. At present, a kind of plastic box is often used in this flower art design.

② Disadvantages of packaging flowers with cartons:

Put in boxes, flowering shoots are not exposed to the air and light so that it is unfavorable for flower preservation.

Cartons are not hard enough to resist extrusion.

Cartons are likely to be damaged by water, so it is necessary to make them water-proof when placing or

（2）纸张包装

包装纸色彩纷呈，花色繁多，在实际选择时要考虑到收花人的性别、年龄及喜好，还要考虑到特定的送花原因，如情人节、母亲节或者圣诞节等节日，需要根据各个节日的文化内涵选择相应的花卉及包装形式。此外，还应考虑到礼物或花卉的大小及质地。

包装纸有各种规格，在包装鲜花时，包装纸的尺寸要剪裁得大小合适，如果包装纸太小，则无法将花卉商品全部包装起来；包装纸太大，又会因折叠太多而使花卉商品显得臃肿、烦琐。

包装纸因材料不同包装的效果也不同。包装鲜花的纸一定要耐揉搓，以便在捆扎处包装纸不会因揉搓而撕裂。包装纸种类繁多，常用的有塑料质地、纸质地、网纱及无纺布等几种。其中，纸质的包装纸发展较快，除普通的印花包装纸外，又出现了皱卷纸、皱纹纸、瓦楞纸、印花纸、云龙纸、彩织纸、彩虹纸、彩钻纸、防水格子纸、云丝纸、手揉纸、麻落水纸、浮染纸、柳染纸、金花纸等。纸质的包装纸质地柔软、纹理自然，给人以返璞归真的美感，而且比较符合环保要求。用纸质的包装纸包装花卉时要注意防潮，以免整个包装变形、变色。

①塑料包装纸：塑料包装纸表面印有精美的图案，色泽鲜明，色彩艳丽，具有光泽度和透明度高，厚薄一致，光滑、不透水、气，不怕水浸的特点，并且价格低廉，是花店最常用的包装材料。塑料包装纸上面印花后，会呈现出透明、不透明、

transplanting cut flowers in cartons.

(2) Paper Packaging

Colorful packaging paper varies in designs and colors. It is necessary to consider receivers' genders, ages and preferences, and the specific reasons of sending flowers in the choice-makings. For example, on such festivals as Valentines' Day, Mother's Day or Christmas, the cultural connotation of different festivals will determine the choice of flowers and their packaging. And furthermore, the size and quality of gifts or of cut flowers should be taken into consideration.

There are various specifications of packaging paper. In packaging flowers, the size of packaging paper should be appropriate. Too small packaging paper will fail to wrap flower products completely; too big packaging paper will make flower products appear clumsy and complicated due to too much folding.

Different materials of packaging paper produce different packaging effects. Flower packaging paper must be rub-resisting so that it will not tear for rubbing in strapping. There is a wide variety of packaging paper. The commonly-used materials involve the qualities of plastic, paper, netted gauze and adhesive-bonded fabric, of which packaging paper with the quality of paper has been more readily and popularly used. Besides ordinary printing packaging paper, there appears crepe rolled paper, crepe rolled paper, corrugated paper, printing paper, Yunlong paper, colored woven paper, rainbow paper, colored jewel paper, waterproof checked paper, Yunsi paper, crumpled paper, linen paper, floating-dyed paper, willow-dyed paper and golden flower paper. Paper packaging has soft and natural texture, gives people natural aesthetic feelings and comparatively meets the needs of environmental protection. In packaging flower products, packaging paper should remain in moisture-proof states in case the whole packaging should get deformed and discolored.

① Plastic packaging paper: Elegant patterns, with bright and lustrous colors, are printed on the surface of

半透明等不同的效果。其中，透明包装可以充分展示鲜花的美丽。塑料包装纸种类丰富。

塑料包装纸因为经过塑化处理，吸湿性较好，受潮后易黏接，所以保存时要防水浸。另外，塑料包装纸具有纵向强度大于横向强度的特点。

②云龙纸：云龙纸的质地柔软，具有极好的韧性，包装鲜花时其硬度不够，纸面容易下垂。而且，花店如果采用普通胶条粘连云龙纸，不容易粘上。因为云龙纸不防水，所以在包装鲜花时，里面最好垫上透明的玻璃纸。云龙纸的色彩柔和、纹理自然，这是其他包装材料所不能比拟的。用云龙纸包装的鲜花显得柔和、精巧、雅致、高贵。

③手揉纸：手揉纸比棉质的质地厚、硬，包装鲜花时纸面易成型，不下垂，且手揉纸颜色亮丽、纹路清晰、耐揉搓，不易撕裂。虽然手揉纸具有一定的抗水性，但在包装鲜花时，纸里面也要垫上玻璃纸防水。用手揉纸包装的鲜花，使人感觉高档、自然。

④皱卷纸：皱卷纸又称褶皱纸，这种纸线条流畅、自然，富含皱褶，极富弹性，易与花的形态融为一体。

plastic packaging paper, which has a high gloss, high transparency and identical thickness, and which is smooth, water-proof and air-proof so that it is not so easily damaged by moisture. Plastic packaging paper is very cheap and thus is the most commonly-used packaging material in flower shops. Printed on its surface, plastic packaging paper can present different effects like being transparent, opaque and translucent. Transparent packaging can fully display the beauty of cut flowers. There is a wide variety of plastic packaging paper.

Through plasticizing processing, plastic packaging paper has better moisture absorption and is easily stuck together after getting wet. Therefore, in storage, it should avoid getting wet. Furthermore, it has the characteristics that its longitudinal strength is greater than its horizontal strength.

② Yunlong paper: On the one hand, with soft texture, Yunlong paper is very flexible. It is not hard enough for packaging flower, so it easily droops. It is also not easy for flower shops to stick Yunlong paper with ordinary adhesive strips. And what's more, because Yunlong paper is not water-proof, it is advised to pad transparent cellophane inside in packaging flowers. On the other hand, Yunlong paper has soft colors and natural texture, which is unparalleled to other packaging materials. Cut flowers packaged with Yunlong paper appear soft, delicate, elegant and noble.

③ Crumpled paper: Crumpled paper is thicker and harder than cotton paper. In packaging flowers, it is easily shaped and doesn't droop. Besides, with clear texture, crumpled paper is brightly colored, rub-resisting and not easy to tear. Although it is water-proof to a certain extent, cellophane should be padded inside crumpled paper in flower packaging. Cut flowers packaged with crumpled paper look high-graded and natural.

④ Crepe rolled paper: Crepe rolled paper is also known as crease paper. This paper has fluent natural lines, full of creases, flexible and easily harmonious with flower patterns.

皱卷纸的长度可拉伸变化，花店在购买时，可以购买纸的边缘已经卷好的皱卷纸。如果边缘未卷好，在没有卷边器的情况下，可以自己动手卷边，方法如下：首先，将皱卷纸的边缘伸拉后放开，使其可用长度加大一些，以节约成本。然后，两只手一上一下朝相反的方向拧一下，这样处理后，纸边缘就会出现不规则的波状花纹。如果想让纸边卷得大一些，也可以两手同时向外卷纸边。

使用卷边皱卷纸包装圆形捧花花束，最基本的方法是用纸将花束围起来，使卷边与花朵高度相同，然后在花束的捆扎点上绑扎收紧。

卷边皱卷纸的花色品种丰富，用它包装的鲜花给人以浪漫、华美的感觉。

卷边皱卷纸的价格较贵，使用前可以稍微将其拉长，以扩大皱褶纸的面积。同时，花里面应垫上玻璃纸，注意防水，有的皱卷纸沾水后，会变软甚至掉色。

（3）网纱包装

包装鲜花的尼龙网纱有国产的，也有进口的，颜色丰富。用网纱包装鲜花，玲珑剔透，使鲜花若隐若现，让人爱不释手，极具美感。

在操作中，可以用订书机使网纱定型。

（4）无纺布

无纺布由纤维构成，具有布的外观，是新一代环保材料，具有轻薄、透气、柔韧、色彩丰富、鲜艳明丽等特点。图案和款式多样，用途广泛，经济实惠。有纯色的和带花纹的。

The length of crepe rolled paper can extended as it is stretched. When florists purchase crepe rolled paper, they can buy crepe rolled paper whose edges have been rolled. If not, in the absence of an edge folder, they can roll the edges on their own in the following ways: First, stretch the edges of crepe rolled paper and then release them to extend its usable length and to save costs. Second, twist them in opposite directions with both hands until there appear irregular wave patterns on the edge of crepe rolled paper. To get bigger edge rolls, florists can roll the edges outward with both hands at the same time.

When florists use edge-rolled crepe rolled paper to package round flower bouquets, the basic method is to wrap bouquets with crepe rolled paper, keep the edges of crepe rolled paper and bouquets at the same level and then bind them up tightly on the strapping point of the bouquets.

There is a great variety of edge-rolled crepe rolled paper in colors and patterns. Cut flowers packaged with it can display their romance and magnificence to consumers.

The price of edge-rolled crepe rolled paper is relatively high, so it can be stretched slightly before use to expand its size. In the meantime, it is advised to pad cellophane inside in packaging flowers to prevent moisture, for some crepe rolled paper may turn soft and even fade in colors after getting wet.

(3) Netted Gauze Packaging

There are domestic and imported colorful nylon netted gauzes for flower product packaging. Partly hidden and partly visible, cut flowers packaged with netted gauze are exquisite, appealing and aesthetic.

In operation, netted gauze packaging materials can fall into a pattern with the help of staples.

(4) Adhesive-bonded Fabric

Made up of fibers, looking like a cloth, adhesive-bonded fabric is a new generation of environmentally-friendly materials and characteristic of being light, thin, ventilating, pliable, colorful and bright. With a variety of patterns and designs, it is widely used, economical and practical, and includes pure and patterned designs.

（5）捆扎材料

捆扎花束的材料很多，有丝带、加入金属丝的丝带、各种形式的装饰丝带、加入花纹的装饰丝带、缎带，有印花的、金色的、银色的等，有透明的、不透明的。选择时要与鲜花包装搭配协调，主要起到固定、陪衬的作用。

（6）其他包装辅助材料

包装时还需要其他一些辅助材料，如可以起到黏合作用的胶枪、胶棒、喷胶、乳胶、乳锅、各色亚龙贴布等，越齐全越好。为了使包装的鲜花更好地适合更多的场合和消费群体，还可以采用一些礼品装饰材料（如小的布娃娃、动物等玩偶）和各种颜色人工合成的彩珠、珠链、石子等。现在花卉市场上还出售一种可降解、可回收利用的物质——聚丙烯颗粒，有瓶装的，也有袋装的。用它装点鲜花，在花朵上面洒满亮晶晶的颗粒，会产生意想不到的效果。

鲜花包装制作时还会用到其他的一些辅助材料，如别针、各色金属线、金属环、玻璃珠、玩具、各种贺卡、心形插、天使发丝等。在插花作品中，将这些物件巧妙构思、搭配，都会使包装好的花卉产生意想不到的效果，增加吸引力。

在花卉商品的包装过程中，大量应用辅助材料，会增加鲜花的美感或者提高插花员包装、制作插花的速度，减少顾客的等候时间。适当开发运用高档包装材料，探索更多的包装方式方法，把花卉包装做得更精美，企业也可以从中获得一定经济效益。

(5) Strapping Materials

There are various kinds of strapping materials for bouquets including ribbons, metal-added ribbons, all kinds of decorative ribbons, patterned decorative ribbons and laces, with a wide variety of choices in colors, designs and effects. For example, there are printed, golden, or silver strapping materials and transparent or opaque strapping materials. The choice of strapping materials depends on flower product packaging. They play a main role in fixing and supplementing the flower product packaging.

(6) Other Supplementary Packaging Materials

Flower product packaging needs other supplementary packaging materials. For example, adhesive materials like adhesive guns, adhesive sticks, adhesive spray, latex, latex pot and various kinds of Yalong adhesive cloth. The more complete the adhesive materials are, the better it is. To better suit packaged flowers to more occasions and consumer groups, some decorative materials like tiny dolls and toy animals, and various synthetic colorful beads, bead chains and stones can also be adopted as supplementary packaging materials. At present, polypropylene granules, a kind of biodegradable and recyclable substance, are on sale in flower markets, bottled or bagged. Making flowers up with them can produce unexpected effects by spreading shiny particles on the flowers.

Other supplementary materials used in flower packaging are pins, all kinds of colorful metal wires, metal rings, glass beads, toys, all kinds of cards, heart-shaped pin holders and angel silky hair. In the works of flower arrangement, ingeniously conceived and matched, these supplementary packaging materials can produce unexpected effects for the packaged flowers and enhance their attraction.

In the process of commodity packaging, widely-applied supplementary packaging materials will enhance the aesthetic feelings, or improve flower arranging staff's efficiency in packaging and arranging cut flowers so as to save customers' waiting time. By properly developing and applying high-graded packaging materials, exploring more packaging methods and producing exquisite packaging,

总之，随着花卉业的发展，鲜花包装材料市场也在不断发展，这必将导致可以用于鲜花包装的材料越来越多。插花人员要不断学习，结合花店的特点挖掘和采用新的包装材料，不断创新，才能在竞争中立于不败之地。

flower business can surely obtain certain economic benefits.

To sum up, with the development of flower business, flower packaging material markets are also increasingly prosperous, which will surely bring about more and more materials suitable for flower packaging. Florists should keep on learning and innovating, develop and adopt new packaging materials so as to be invincible in the competition.

参 考 文 献

长谷惠. 2010. 心意满满的节日鲜花包装平装[M]. 朱宁，译. 郑州：河南科技出版社.

高俊平. 2002. 观赏植物采后生理与技术[M]. 北京:中国农业大学出版社.

赖尔聪. 2000. 昆明世博会插花大赛获奖作品集[M]. 合肥：安徽科学技术出版社.

劳动和社会保障部教材办公室. 2004. 插花员(中级) [M]. 北京：中国劳动社会保障出版社.

劳动和社会保障部教材办公室. 2007. 插花员(高级) [M]. 北京：中国劳动社会保障出版社.

卢正言. 2003. 花趣[M]. 上海:学林出版社.

田蓉蓉. 2005. 韩式花卉包装[M]. 郑州:河南科技出版社.

王莲英．秦魁杰．尚纪平. 1998. 插花创作与赏析[M]. 北京：金盾出版社.

吴红芝，赵燕. 2012. 鲜切花综合保鲜技术与疑难解答[M] . 北京：中国农业出版社.

犀文图书. 2012. 花束包装技法 [M]. 长沙：湖南美术出版社.

徐海宾. 1996. 赏花指南[M]. 北京：中国农业出版社.

张颢，王继华，唐开学，等. 2009. 鲜切花：实用保鲜技术[M]. 北京：化学工业出版社.

周武忠，陈筱燕. 1999. 花与中国文化[M]. 北京：中国农业出版社.

附录　插花艺术术语英汉对照表

中文	英文	中文	英文
插花艺术	flower arrangement art	婚庆插花	wedding flower arrangement
花艺	floriculture	会场插花	flower arrangement in meeting place
佛教供花	flower arranging for Buddhism	写字楼插花	flower arrangement in office building
理念花	concept flower	商厦插花	flower arrangement in commercial building/mall
心象花	mood flower (xinxianghua)		
瓶花	vase flower arrangement	丧用插花	flower arrangement for funeral affairs
地坊流	Ikenobo	新娘花	decorative flower for bride
立华	Tachibana	捧花	bouquet
袁宏道流（宏道流）	Yuan Hongdao school (Hongdao School)	花托	receptacle
		头花	headdress flower
剑山	tsurugi-san	胸花	brooch
池坊流	Ikebana	腕花	wrist flower
小原流	Ohara	花车	float
草月流	Sogetsu	车头花	front flower
日本花道	Japan ikebana	车门花	door flower
东方插花	oriental flower arrangement	镶边花	embroidered flower
西方插花	western flower arrangement	撒花	strewing flower
古代插花	ancient flower arrangement	线状花材	lineal flower material
宗教插花	religious flower arrangement	团块状花材	crumby/cloddy flower material
宫廷插花	court flower arrangement/ flower arrangement at royal court	填充花	filling flowers
		主花	the main parts of flower arrangement
民间插花	folk flower arrangement	草本切花花材	herbal cutting flower material
文人插花	flower arrangement by scholar	观花花材	flower material for appreciating flowers
现代插花	modern flower arrangement	观叶花材	flower material for appreciating leaves
礼仪插花	etiquette flower arrangement	木本切花花材	woody cutting flower material
大众插花	public flower arrangement	观果花材	flower material for appreciating fruits
艺术插花	artistic flower arrangement	插花器皿	equipment for flower arrangement
现代花艺设计	modern floral design	瓶花	flowers in vase
写景插花	flower arrangement in scenery	盘花	flowers in plate
写意插花	freehand flower arrangement	缸花	flowers in crock
抽象插花	abstract flower arrangement	碗花	flowers in bowl
家庭插花	family flower arrangement	筒花	flowers in tube
宾馆插花	hotel flower arrangement	篮花	flowers in basket

花泥	flower mud	皱卷纸	crepe rolled paper, crease paper
花插	flower rece ptacle	皱纹纸	crepe paper
花胶	flower glue	瓦楞纸	corrugated paper
垫座	pedestal	印花纸	printing paper
去刺	remove thorn	云龙纸	yunlong paper
剥花	stripping	彩织纸	colored woven paper
圈叶	circling	彩虹纸	rainbow paper
翻翘	warping	彩钻纸	colored jewel paper
焦点花	focal flower	防水格子纸	waterproof checked paper
块状花材	clumpy floral material	云丝纸	yunsi paper
线状花材	linear floral material	手揉纸	crumpled paper
花德	flower virtue	麻落水纸	linen paper
花卉欣赏	flower appreciation	浮染纸	floating-dyed paper
花文化	flower culture	柳染纸	willow-dyed paper
花语	flower language	金花纸	golden flower paper
盆花	potted flower	塑化处理	plasticizing processing
花冠	corollas	玻璃纸	cellophane
花瓣	pedal	手揉纸	crumpled paper
苞片	bract	棉纸	cotton paper
小花	floret	卷边器	edge folder
花卉包装	flower packaging	圆形捧花花束	round flower bouquets
透明式包装	transparent packaging	丝带	ribbon
半透明式包装	translucent packaging	缎带	lace
塑料包装	plastic packaging	胶枪	adhesive gun
纸包装	paper packaging	胶棒	adhesive stick
纸盒包装	carton packaging	喷胶	adhesive spray
尼龙纱网包装	nylon gauze packaging	乳胶	latex
礼盒包装	box packaging	乳锅	latex pot
包装规格	specifications of packaging	亚龙贴布	yalong adhesive cloth
网纱	netted gauze	插花作品	the works of flower arrangement
无纺布	adhesive-bonded fabric	插花员	flower arranging staff
印花包装纸	printing packaging paper		

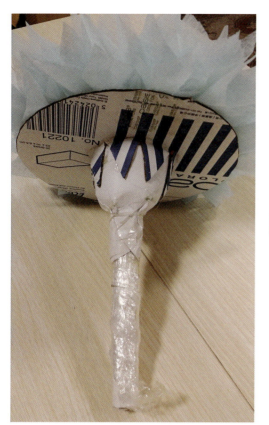

图3-1 捧花托	图3-2 新娘捧花
Figure 3-1 Holding receptacle	Figure 3-2 Bride bouquet
图3-3 新娘头花	图3-4 花 车
Figure 3-3 Bride's headdress flower	Figure 3-4 Float

图3-5 花泥的固定
Figure 3-5 fixation of flower mud

图3-6 花 车
Figure 3-6 Float

图3-7 花车车链
Figure 3-7 Float Flower chain

Color Illustration

图3-8 胸花设计 Figure 3-8 Brooch design	图3-9 胸花的固定 Figure 3-9 Fixation of brooch	图3-10 庆典花篮 Figure 3-10 Celebration flower baskets
图3-11 四面观庆典花篮 Figure 3-11 All-side celebration flower basket	图3-12 花圈 Figure 3-12 Wreaths	

图3-13 会议桌花
Figure 3-13 Board flower

图3-14 下垂形会议桌花
Figure 3-14 Drooping board flower

图3-15 以非洲菊为主要花材的插花
Figure 3-15 Flower arrangement with gerbera as main floral material

图3-16 香石竹为主的花篮
Figure 3-16　Flower basket with carnation as main floral material

图3-17　花　　束	图3-18　生日花束
Figure 3-17　Bouquet	Figure 3-18　Birthday bouquet

图3-19 彩菊花篮
Figure 3-19 Basket of colored chrysanthemums

图4-1 《雷雨》（杨发顺摄）
Figure 4-1 *Thunderstorm* (Photographed by Yang Fashun)

图4-2 《百折不挠》（杨发顺摄）
Figure 4-2 *Perseverance* (Photographed by Yang Fashun)

图4-3 《君子之交》（杨发顺摄）
Figure 4-3 *Friendship between gentlemen*
(Photographed by Yang Fashun)

图4-4 《天音》（杨发顺摄）
Figure 4-4 *Music of heaven*
(Photographed by Yang Fashun)

正面

俯视

背面

图4-5 《喜见恩师》(郑丽摄)
Figure 4-5 *A happy reunion with the teacher* (Photographed by Zheng Li)

图4-6 《千古》（艾万峰摄）
Figure 4-6 *Immortality* (Photographed by Ai Wanfeng)

图5-1 串 联
Figure 5-1 Connecting

图5-2 堆 叠
Figure 5-2 Stacking

图5-3 平行线
Figure 5-3 Parallel lines

图5-4 点、线、面、体综合立体构成
Figure 5-4 Comprehensive constitution of point, line, surface and body

图5-5 支架
Figure 5-5　Scaffolding

图5-6 交叉线
Figure 5-6　Crossing Lines

图5-7 面材的立体构成
Figure 5-7　Three-dimensional composition of panels

图5-8 体的立体构成
Figure 5-8 Three-dimensional composition of body

图5-9 表现个体花材的特质
Figure 5-9 Demonstrating the nature of a particular floral material

图5-10 花材布局聚中有散、静中有动
Figure 5-10 Creating the floral arrangement of scattering mingled with gathering and the static mingled with the dynamic

图5-11 枝的一枝突出
Figure 5-11 Highlighting-one-branch in branch arrangement

图5-12 叶的一枝突出
Figure 5-12 Highlighting-one-branch in leaf arrangement

图5-13 花的一枝突出
Figure 5-13 Highlighting-one-branch in flower arrangement

图5-14 平中出奇
Figure 5-14 The extraordinary out of the ordinary

图5-15 对角平衡
Figure 5-15 The balance of opposite angles

图5-16 造型的斜角呼应
Figure 5-16　Oblique angle echoing in shaping

图5-17 色彩的斜角呼应
Figure 5-17　Oblique angle echoing in color

图5-18 组群形式
Figure 5-18　Forms of grouping

图5-19 花材多而不杂、变而不乱
Figure 5-19　Floral materials being rich but not mussy, and changeable but not messy

图5-20 按花材不同品种分组组合
Figure 5-20 Grouping according to types of floral materials

图5-21 按花材不同色彩分组组合
Figure 5-21 Grouping according to colors of floral materials